Regional Climate Change and Variability

NEW HORIZONS IN REGIONAL SCIENCE

Series Editor: Philip McCann, *Professor of Economics, University of Waikato, New Zealand and Professor of Urban and Regional Economics, University of Reading, UK*

Regional science analyses important issues surrounding the growth and development of urban and regional systems and is emerging as a major social science discipline. This series provides an invaluable forum for the publication of high quality scholarly work on urban and regional studies, industrial location economics, transport systems, economic geography and networks.

New Horizons in Regional Science aims to publish the best work by economists, geographers, urban and regional planners and other researchers from throughout the world. It is intended to serve a wide readership including academics, students and policymakers.

Titles in the series include:

Industrial Clusters and Inter-Firm Networks
Edited by Charlie Karlsson, Börje Johansson and Roger R. Stough

Regions, Land Consumption and Sustainable Growth
Assessing the Impact of the Public and Private Sectors
Edited by Oedzge Atzema, Piet Rietveld and Daniel Shefer

Spatial Dynamics, Networks and Modelling
Edited by Aura Reggiani and Peter Nijkamp

Entrepreneurship, Investment and Spatial Dynamics
Lessons and Implications for an Enlarged EU
Edited by Peter Nijkamp, Ronald L. Moomaw and Iulia Traistaru-Siedschlag

Regional Climate Change and Variability
Impacts and Responses
Edited by Matthias Ruth, Kieran Donaghy and Paul Kirshen

Industrial Agglomeration and New Technologies
A Global Perspective
Edited by Masatsugu Tsuji, Emanuele Giovannetti and Mitsuhiro Kagami

Regional Climate Change and Variability

Impacts and Responses

Edited by

Matthias Ruth
University of Maryland, College Park, Maryland, USA

Kieran Donaghy
University of Illinois, Champaign, Illinois, USA

Paul Kirshen
Tufts University, Medford, Massachusetts, USA

NEW HORIZONS IN REGIONAL SCIENCE

Edward Elgar
Cheltenham, UK • Northampton, MA, USA

Published by
Edward Elgar Publishing Limited
Glensanda House
Montpellier Parade
Cheltenham
Glos GL50 1UA
UK

Edward Elgar Publishing, Inc.
136 West Street
Suite 202
Northampton
Massachusetts 01060
USA

A catalogue record for this book
is available from the British Library

Library of Congress Cataloguing in Publication Data

Regional climate change and variability : impacts and responses / edited by
Matthias Ruth, Kieran Donaghy, Paul Kirshen.
    p. cm. – (New horizons in regional science series)
    Includes bibliographical references and index.
    1. Climatic changes–Environmental aspects–Evaluation. 2. Environmental
impact analysis. I. Ruth, Matthias. II. Donaghy, Kieran. III. Kirshen, Paul
H. IV, New horizons in regional science.
    QC981.8.C5R415 2006
    333.71'4–dc22

                                                          2005033749

ISBN-13: 978 1 84542 599 9
ISBN-10: 1 84542 599 5

Printed and bound in Great Britain by MPG Books Ltd, Bodmin, Cornwall

Contents

List of Contributors

William Anderson is the Associate Chairman and Professor of the Department of Geography and the Center for Transportation Studies at Boston University. His research interests include economic geography, transportation studies, urban geography, energy and environmental studies, urban and regional economic modeling, interregional and international migration, international trade, and quantitative methods. He is also the editor for the *Energy Studies Review* and until 1998, was the editor of the *Canadian Journal of Regional Science*.

John Antle is Professor of Agricultural Economics and Economics at Montana State University. He is the principal investigator for the USAID Soil Management Collaborative Research Support Program. His research interests include soil management, soil mitigation, and greenhouse gases and environmental impacts on agriculture. From 1999–2000, Dr Antle was the president of the American Agricultural Economics Association and was named as a distinguished fellow by the Association in 2002. He was a senior economist on the President's council of economic advisors from 1989–1990. He has extensively published on topics ranging from soil carbon sequestration, to agroecosystems, to spatial heterogeneity.

Levi Brekke's current work is focused on reservoir operations modeling, hydrologic studies, decision analysis, and applications for climate information in Reclamation's short- and long-term planning processes (ranging from seasonal outlooks to multidecadal futures). Dr Brekke is Reclamation's representative on a joint agency Climate Change Work Team with the California Department of Water Resources. Work efforts are focused on developing climate change risk information for State and Federal infrastructure planners. Dr Brekke's previous work includes climate change impacts assessment for California's Central Valley water resources and Lake Cachuma water quality, climate teleconnection applications for California reservoir operations, and various projects in water and wastewater treatment consulting.

Susan Capalbo is Professor and Director of Special Projects for Enhancing Diversity at Montana State University. Her research interests include environmental–economic tradeoff analysis, climate change and carbon sequestration, productivity analysis and production theory, and rural health care and cost–benefit analysis. She teaches courses in environmental economics, natural resources valuation, and production and development economics.

Gary Christopherson is the Director of the Center for Applied Spatial Analysis and an Adjunct Research Social Scientist for the Department of Geography and Regional Development at the University of Arizona. Christopherson is also a PhD candidate in Near East Studies and a Research Assistant in the Advanced Resource Technology Group at the University of Arizona.

Kevin Civerolo has been a research scientist with the New York State Department of Environmental Conservation Division of Air Resources since 1998. Dr Civerolo's primary task is to provide technical support for the state planning process for ozone, fine particulates and mercury. His professional interests include the evaluation of meteorological and photochemical models; estimating the effects of land use change and large-scale tree planting on air quality; analysis of spatial and temporal trends in air and water pollution data using traditional and non-traditional methods; and back trajectory and source attribution analysis. Dr Civerolo also has experience in the development and use of several techniques for monitoring ambient reactive nitrogen compounds. He currently is an adjunct assistant professor at the University at Albany School of Public Health. His MS and PhD degrees in Meteorology were awarded by the University of Maryland in 1993 and 1996, respectively.

Mike Crimmins is a PhD candidate in the Department of Geography and Regional Development at the University of Arizona. His research interests include land–atmosphere interactions and the impact of landscape changes on local and regional climates, as well as the spatial relationships between terrestrial ecosystems and the atmospheric environment, environmental decision making, and ecologically relevant engineering design.

Kieran Donaghy is Associate Professor of Urban and Regional Planning at the University of Illinois at Urbana-Champaign. The thread connecting Dr Donaghy's diverse areas of research, teaching and practice is an interest in developing dynamic models that help communities, states and countries understand how complex social and natural systems are interrelated and how different courses of action may alter evolving patterns for better or worse. His applied research focuses on regional impacts of climate change, transportation planning, state and local development, macroeconomic policy, arms race and military spending issues, and the coordination of international economic policies.

John Dracup has been a professor in Civil and Environmental Engineering at Berkeley since June 2000. Prior to that he was a professor in the equivalent department at UCLA. At Berkeley he teaches courses on 'Design of Environmental and Water Resource Systems,' Fluid Mechanics, Water Resources Planning and Management, and International Water Systems. The focus of his research program is in hydrology and water resource systems analysis. In the area of hydrology he has been involved in the stochastic analysis of floods and droughts and the assessment of the impact of climate on hydrologic processes. In the area of water resources his research interests are in the simulation and optimization of groundwater systems and large-scale river basin systems. He has been a Principal Investigator for research grants from the United Nations Development Program, the National Science Foundation, the Ford Foundation, the Office of Naval Research, the Environmental Protection Agency, the Office of Water Resources Research, the California Air Resources Board, the Metropolitan Water District of Southern California, the UC Water Resources Center, the UC Pacific Rim Research Center, and the National Institute for Water Resources Research. He was the Senior Fulbright Scholar to Australia in 2000.

J. Wayland Eheart is Professor of Environmental Systems in the Department of Civil and Environmental Engineering at the University of Illinois, Urbana-Champaign. Dr Eheart's research focuses on the evaluation and development of policies, rules and regulations, and, in particular, on the interface of those policies, rules and regulations with engineering decision making. This work involves the use of models of engineered systems coupled with models of natural systems and principles of economic decision making. Dr Eheart teaches courses in environmental systems, uncertainty and risk analysis and water quality modeling.

Richard Goldberg is a research staff associate at the Goddard Institute for Space Studies, Columbia University Center for Climate Systems, New York, NY. He received his BS (1979) and MS (1982) in meteorology from the City College of New York.

Edwin Herricks is a faculty member in the Department of Civil and Environmental Engineering at the University of Illinois, Urbana-Champaign. He received a BA in Zoology and English from the University of Kansas, an MS in Sanitary/Environmental Engineering from The Johns Hopkins University, and his PhD in Biology from Virginia Polytechnic Institute and

State University. Dr Herricks has been an advisor to local, state and federal government, and has regularly served as an advisor to the US Environmental Protection Agency. His recent research includes evaluation of the regional effects of climate change, specifically the effect of climate change scenarios on fisheries; the restoration of streams in urban areas, including the development of ecological engineering concepts for watershed management; development of an integrated hydrologic, geomorphic and ecological classification system for watershed management, and the development of a systems approach to minimizing wildlife/aircraft interactions and improving aircraft safety.

Hugo Hidalgo is a researcher for the Climate Research Division at Scripps Institution of Oceanography. His research interests are related to hydroclimatology, surface water hydrology, paleoclimate and climate change. His current research is related to the variability and change of hydroclimatological parameters in the Western United States that are indicators of aridity and drought at a variety of temporal scales.

Christian Hogrefe is an Adjunct Research Associate at the Atmospheric Sciences Research Center, University at Albany. Dr Hogrefe received a PhD and MS in Atmospheric Sciences from the University at Albany. He currently is working with Dr Kinney to simulate ozone air quality under present and future regional climate scenarios as part of the New York Climate and Health Project. Dr Hogrefe also has extensive experience in the development and application of evaluation techniques to assess the performance of meteorological and photochemical modeling systems, in the use of photochemical models for real-time air quality forecasting, and in the integration of air quality observations and model predictions in the regulatory framework.

John Keyantash is an Assistant Professor in the Department of Earth Sciences at California State University, Dominguez Hills. He joined the department in Fall 2002. In addition to teaching upper division science courses for geography and geology majors, as well as general education students, Dr Keyantash performs research on droughts and other trends in the hydrologic cycle. He is involved in the development of classroom science units for the Los Angeles Unified School District. Prior to joining California State University, he was a postdoctoral fellow at UC Berkeley, researching the potential effects of climate change upon chinook salmon populations. He graduated from the University of California, Los Angeles (UCLA) in 2001 with a PhD in Civil Engineering, with an emphasis in Water Resources Engineering. He also holds a master's degree in Atmospheric Sciences from

UCLA, and obtained his bachelor's degree in Environmental Science (General Science) at Oregon State University.

Patrick Kinney is an Associate Professor at the Mailman School of Public Health (MSPH) at Columbia University. He has carried out numerous studies examining the human health effects of air pollution, including studies of the effects of ozone and/or particulate matter on lung health and on daily mortality in large cities. His recent research has focused on characterizing levels and determinants of indoor, outdoor and personal exposures to air pollution in the underprivileged neighborhoods of NYC, including studies of indoor allergens, diesel vehicle emissions, volatile organic compounds, PAHs and other air toxins. In 2001, Dr Kinney established a new research program at MSPH to develop and apply integrated models for assessing the human health impacts of climate variability and change. He directs the New York Climate and Health Project, an interdisciplinary team of health, air quality and climate scientists examining changes in health impacts related to heat stress and air pollution in the coming century due to variations in climate and land use in the NYC metropolitan area.

Paul Kirshen is Research Professor in the Civil and Environmental Engineering Department at Tufts University, as well as the Director of Tufts Water, Sustainability, Health and Ecological Diversity (WaterSHED) Center and the Co-founder and Steering Committee Member of The Mystic Watershed Collaborative. His expertise includes water resources planning and management, integrated assessment, climate change, water policy analysis and hydrology. He conducts research in developed and developing countries on climate change impacts and adaptation of integrated water resources and watershed planning; management; and policy.

Kim Knowlton is a post-doctoral research scientist at the Mailman School of Public Health, Columbia University. Her dissertation research was conducted with the NY Climate and Health Project on projecting near-future public health impacts of climate, air quality, land use and demographic changes within the New York region. She holds an MS in Environmental and Occupational Health Sciences from Hunter College, City University of NY, and a BA in Geological Sciences from Cornell University. Her past projects include administration of a national medical screening of nuclear workers; tracking occupational injuries and illnesses among hospital workers (in conjunction with the US Centers for Disease Control and Prevention); evaluation of environmental health and safety concerns for proposed radioactive waste facilities; review of closure plans for Fresh Kills landfill on Staten Island (for the New York City Public Advocate's Office); and re-evaluation of chemical exposures from fish consumption (for the Natural Resources Defense Council).

Jia-Yeong Ku received professional training from the Department of Atmospheric Science, State University of New York and obtained his PhD in 1984. He joined the Department of Environmental Conservation of New York State as a research scientist in 1985. Dr Ku works on the modeling and understanding of fundamental processes taking place in the formation of ozone and PM episode in the eastern US. His research interests include the development and improvement of the planetary boundary layer process in air quality numerical models and modeling results data analysis.

Barry Lynn Lynn received a BA from Oberlin College in 1986, a MS from Pennsylvania State University in 1988, and a PhD from Rutgers University in 1994, studying biology, meteorology and environment. His most recent work involves coupling a three-dimensional atmospheric mesoscale model (MM5) with the Goddard Institute for Space Studies climate prediction model (GISS GCM) to simulate future climate variability on a local scale for the New York Climate and Health Project. Dr Lynn's other projects involve coupling MM5 to spectral microphysics. The new model has led to a significant improvement in the simulation of precipitation processes. Dr Lynn worked under the direction of Dr Rosenzweig to develop fine-scale gridded maps of hourly surface meteorology over NY State from 1988 to 2002 (15 years) using station observations integrated with the MM5.

Norman Miller is a Hydrometeorologist at the University of California's Berkeley National Laboratory and is an Adjunct Professor in the Department of Hydrology and Water Resources at the University of Arizona–Tucson. He leads the Atmosphere and Ocean Sciences Group at Berkeley Lab. His research includes analyzing atmosphere and land surface processes at a range of scales, evaluating climate change impacts, and advancing new computational techniques for climate simulations. He has published over 50 peer reviewed journal papers and book chapters, is a contributing author of the Intergovernmental Panel for Climate Change Second and Third Assessment Reports, the Southwestern US National Assessment, and the California Assessment Reports.

Barbara Morehouse's research emphasizes institutional and policy analysis in the framework of natural resource management and environmental change. She brings a background in critical theory and cultural studies to her research activities, as well as empirical research experience in the Southwest and US–Mexico border region. Currently, Dr Morehouse is carrying out institutional analyses of climate impacts on water management in the Southwest and of complex society–environment relationships influencing fire management in the region. She also sustains an ongoing interest in the roles played by boundaries and border areas in science–society contexts.

Charles Oliveri has worked extensively as a computer programmer for academic and business applications. His expertise is in C++ and related programs. Currently he is a graduate student at SUNY Brockport in the Computer Science Department and lives in Syracuse, NY.

Brian Orland is Head of Landscape Architecture at Penn State University. From 1982–2000 he was Professor of Landscape Architecture and Director of the Imaging Systems Laboratory, at the University of Illinois. He has taught design and land resource evaluation with particular emphasis on human–environment interactions and environmental perception. His research interests include the computer modeling of environmental impacts and the design of online information systems to support community-based planning initiatives. Studies have included the impacts of highway development, insect pest impacts and logging on national forests, and military training activities.

Barron Orr is Assistant Professor and Geospatial-extension Specialist at the University of Arizona's college of Agriculture. His research interests include geospatial technology, remote sensing, natural resource management, land tenure, land degradation and diffusion of innovation. His primary objective is to join the missions of the NASA Office of Earth Science and Space Grant with the experience and infrastructure of the Cooperative Extension in order to bridge the gap between geospatial technology and its potential users in the state of Arizona. His extension programming involves a precision approach to natural resource management and agriculture. Dr Orr is also currently serving as the Associate Director for the UA/NASA Space Grant Program.

Jonathan Overpeck is Professor in the Geosciences Department at the University of Arizona and Director of the Institute for the Study of Planet Earth, as well as the liaison for acquisition and analysis of climate data and project advisor. His research focuses on global change dynamics, particularly developing the regional to global paleoclimatic and paleoceanographic perspectives needed to reconstruct and understand the full range of climate system variability; recognize and anticipate possible 'surprise' behavior in the climate system; understand how the earth system responds to changes in climate forcing; evaluate the realism of environmental models, in particular how they simulate the response to altered forcing; and detect and attribute environmental change to various natural and non-natural forcing mechanisms.

Keith Paustian is a senior research scientist for the Natural Resource Ecology Laboratory and is a professor in the Department of Soil and Crop Science at Colorado State University. Dr Paustian's main fields of interest include agroecosystem ecology, soil organic matter dynamics and global change. His current research includes studies of the mechanisms controling soil carbon dynamics in managed ecosystems, regional and national assessments of agricultural practices to mitigate greenhouse gas emissions, and development of decision support tools to advise farmers, land managers and policy makers on greenhouse gas mitigation strategies.

Nigel Quinn is the Leader of the HydroEcological Engineering Advanced Decision Support Group at Berkeley National Laboratory, is an Adjunct Research Professor of Plant Science at Fresno State University and a Research Engineer in the Civil and Environmental Engineering Department at the University of California, Merced. His research interests are at the interface of hydrology, ecology and environmatics – specifically the development of mathematical models and other decision support tools to improve water and environmental management in California and around the world. Dr Quinn has worked on irrigation- and drainage-related water quality problems in the San Joaquin Valley of California for the past 20 years and is the author of over 50 scientific publications related to soil and water management in the United States, England and Africa.

Joyce Rosenthal is an environmental planner and health scientist with over twenty years of experience in research and community-based environmental improvement projects. She is presently Project Manager for the New York Climate and Health Project, providing oversight and coordination of multi-disciplinary research teams working to establish an integrated modeling assessment of the public health impacts of climate and land use variability in the New York metropolitan region. Ms Rosenthal has previously served as Assistant Director of the Mayor's Council on the Environment of New York City (CENYC), coordinating interagency policy and public outreach on urban environmental issues. She is also Adjunct Professor of Urban Planning at Columbia University's Graduate School of Architecture, Planning and Preservation. Ms Rosenthal received an MS in Urban Planning (2000) and an MPH in Environmental Health Sciences from the Mailman School of Public Health (2001).

Cynthia Rosenzweig directs research focused on the impacts of environmental change, including increasing carbon dioxide, global warming and the El Niño–Southern Oscillation, on regional, national and global scales. She is the Leader of the Climate Impacts Group at the Goddard Institute for Space Studies, and was the Co-leader of the Metropolitan East Coast Regional

Assessment of Climate Variability and Change. She is a Lead Author of the Intergovernmental Panel on Climate Change Working Group II Third Assessment Report.

Matthias Ruth is the Roy F. Weston Chair in Natural Economics, Director of the Environmental Policy Program at the School of Public Policy and Co-director of the Engineering and Public Policy Program at the University of Maryland. His research focuses on dynamic modeling of non-renewable and renewable resource use, industrial and infrastructure systems analysis, and environmental economics and policy. His theoretical work heavily draws on concepts from engineering, economics and ecology, while his applied research utilizes methods of non-linear dynamic modeling and adaptive management. He collaborates extensively with scientists and policy makers in the United States, Canada, Europe, Asia and Africa. Professor Ruth teaches nationally and internationally courses and seminars on economic geography, micro-economics and policy analysis, ecological economics, industrial ecology and dynamic modeling at the undergraduate, graduate and PhD levels, and on occasion conducts short courses for decision makers in industry and policy.

Joel Scheraga is National Program Director for the Global Change Research Program and the Mercury Research Program at the US Environmental Protection Agency (EPA). He directs assessments of the potential impacts of global change on air quality, water quality, ecosystems and human health. Dr Scheraga is also the EPA Principal Representative to the US Climate Change Science Program (CCSP), which coordinates and integrates scientific research on climate and global change supported by the US Government. He was a Lead Author of the 1997 Intergovernmental Panel on Climate Change (IPCC) North American Regional Assessment, and an Assisting Lead Author for the 1994 IPCC Technical Guidelines for Assessing Climate Change Impacts and Adaptations. He holds degrees in geology-mathematics/physics and economics from Brown University.

Chris Small is a geophysicist at the Lamont-Doherty Earth Observatory of Columbia University. His current research interests include the use of satellite remote sensing to quantify changes in the earth's surface and the causes and consequences of these changes. His contribution to the New York Climate Change Project (NYCHP) is the production of the satellite-derived maps of vegetation fraction and albedo used to categorize urban and suburban land cover properties for the climate and air quality models. Chris received a PhD in Geophysics from the Scripps Institution of Oceanography at the University of California San Diego in 1993.

William Solecki's research focuses on the urban environmental change and urban land use and suburbanization. Dr Solecki has recently served on the US National Research Council, Special Committee on Problems in the Environment (SCOPE). He has served as the co-leader of several climate impacts and land use studies in the New York metropolitan region, including the Metropolitan East Coast Assessment of Impacts of Potential Climate Variability and Change. He currently is a member of the International Geographical Union (IGU) Megacity Study Group and the International Human Dimensions Programme on Global Environmental Change (IHDP), Urban Environmental Change Study Group, and is on the editorial board of three journals, *Professional Geographer*, *Urban Ecosystems* and *Social Science Quarterly*. He recently served as the Chair of the Human Dimensions of Global Change Specialty Group of the Association of American Geographers. Solecki is a Professor of Geography at Hunter College – City University of New York. Dr Solecki received his PhD in Geography from Rutgers University.

Thomas Swetnam is Director of the Tree Ring Laboratory; Professor of Dendrochronology; Associate Professor of Watershed Management and Adjunct Associate Professor of Geography and Regional Development at the University of Arizona. His research interests include disturbance ecology, forest ecology, dendrochronology, landscape ecology, interactions of climate, people and ecosystems at time scales of seasons to millennia, spatial scales of forest stands to landscapes, and applications of ecosystem and environmental sciences to land management. In 2002, he received the Henry Cowles Award from the Association of American Geographers for biogeography specialty group. He is the Program Chair for the Ecological Society of America and the Associate Editor of the *International Journal of Wildland Fire*.

Stephen Yool is Associate Professor of Geography and Regional Development, interim department head and an Adjunct Associate Professor of Planning at the University of Arizona. His research interests include biogeography, remote sensing and geographic information systems, with particular applications to the study of fire (pyrogeography), disturbance and disease. He is on the editorial board of *Photogrammetric Engineering and Remote Sensing*.

Acknowledgment

This volume is the product of truly collaborative efforts by its many contributors, their colleagues and friends. Special recognition goes to Dana Coelho at the University of Maryland's School of Public Policy for her expediency in bringing the various chapters into shape, and Tara Gorvine at Edward Elgar's Massachusetts office, who kept her keen eye on both the big picture and fine details. Last but not least, our personal thanks and appreciation go to our families for their love and patience throughout the years.

Foreword

Joel Scheraga

Climate change is real. The climate has changed, is changing, and will continue to change, regardless of any human influence. But since the Industrial Revolution, human activities have begun to significantly affect the Earth's atmosphere and climate, and the changes are expected to continue for the foreseeable future. Although the timing and magnitude of future climate change is uncertain, it will have consequences for human health, ecosystems, economic activity and social well-being. Some of the effects will be harmful, and some beneficial.

The significance of the climate change issue was captured in the Joint G8 Statement, issued on 8 July 2005:

> Climate change is a serious and long-term challenge that has the potential to affect every part of the globe. We know that increased need and use of energy from fossil fuels, and other human activities, contribute in large part to increases in greenhouse gases associated with the warming of our Earth's surface. While uncertainties remain in our understanding of climate science, we know enough to act now to put ourselves on a path to slow and, as the science justifies, stop and then reverse the growth of greenhouse gases.

There are two approaches for dealing with climate change. One strategy is to mitigate the emission of gases that contribute to warming – the so-called 'greenhouse gases' (GHGs). Since the Industrial Revolution, human activities, particularly the burning of fossil fuels, have contributed to increases in the atmospheric concentrations of CO_2, one of the more significant and long-lived greenhouse gases, from about 280 ppm to 377 ppm in 2004 (Keeling and Whorf 2005; IPCC 2001). The mitigation of GHG emissions provides a mechanism for slowing the buildup of GHGs in the atmosphere and the rate of climate change. It is important to mitigate because the rate of climate change may be of greater concern than the magnitude of change for many systems, particularly ecological systems.

A second strategy is to adapt in anticipation of future climate change. Adaptive actions are those responses or actions taken to enhance the resilience of systems sensitive to changes in climate, thereby reducing the risks and taking advantage of the opportunities presented by climate change (National Academy of Sciences 1992). It is essential to adapt because the climate will continue to change regardless of actions taken to mitigate GHG emissions. There is a lag in the time it takes the climate system to respond to changes in atmospheric concentrations of GHGs, so past GHG emissions have already committed us to some amount of future climate change. Anticipatory adaptation may also have the side benefit of increasing the resilience of various systems to natural variations in the Earth's climate system. A sensible policy package will consist of both mitigation and adaptation strategies.

An understanding of the potential consequences of climate variability and change is essential for the development of both mitigation and adaptation strategies. Decisions about the appropriate magnitude and timing of actions to mitigate GHG emissions will depend, in part, on the magnitude of expected impacts (the so-called 'consequences of inaction'). Sensible adaptation strategies cannot be developed until one first understands the sensitivity of systems to changes in climatic conditions and the anticipated impacts that may warrant an adaptive response. But assessment of system sensitivity and anticipated impacts is a complex undertaking. Since changes in climate vary by location, there will be a regional 'texture' to the impacts of climate change. Adaptation strategies must therefore be site-specific. Further, there will be distributional effects across demographic groups as well as across geographic regions. Affected populations will vary, depending upon the effect of climate change being considered. One person's risk may be another person's opportunity. For example, decreases in wintertime snowfall in the Great Lakes region may hurt the skiing industry, but would also reduce the costs of snow removal to different communities. Hence, at any particular location, the distributional effects of climate change must also be assessed, to inform decision makers about the tradeoffs they may have to make as they choose among different adaptation strategies. Decisions about the investment of scarce resources in adaptive responses will inherently be value laden, and decision makers will have to represent the values of their communities in the tradeoffs they make (Scheraga and Grambsch 1998).

This book reports on work that provides new insights about the potential consequences of climate change, and possible adaptation strategies in particular places. The work was sponsored by the US Environmental Protection Agency's (EPA's) Global Change Research Program within the Office of Research and Development (ORD).

The EPA's Global Change Research Program has its primary emphasis on evaluating the potential consequences of climate variability and change on air quality, water quality, ecosystems and human health in the United States. This includes improving the scientific basis for evaluating effects of global change in the context of other stressors, and evaluating the risks and opportunities presented by global change. The EPA uses the results of these studies to investigate adaptation options to improve society's ability to effectively respond to the risks and opportunities presented by global change. The program is multidisciplinary and emphasizes the integration of the concepts, methods and results of the physical, biological, and social sciences into decision support frameworks. As called for by the National Research Council (2001), the EPA supports and fosters projects that link knowledge producers and users in a dialogue that builds a mutual understanding of what is needed, what can credibly be said, and how it can be said in a way that maintains scientific credibility.

The work done by the EPA's program is coordinated and consistent with the 2003 Strategic Plan of the US Climate Change Science Program (CCSP 2003). The CCSP coordinates and integrates scientific research on global change and climate change sponsored by 13 participating departments and agencies of the US Government. The CCSP incorporates the US Global Change Research Program, established by the Global Change Research Act of 1990, and the Climate Change Research Initiative, established by President Bush in 2001. The EPA coordinates with other CCSP agencies to develop and provide useful and scientifically sound information in a timely fashion to decision makers.

The CCSP Strategic Plan calls for the development of information resources to support adaptive management and planning for responding to climate variability and change. The Plan calls for research that integrates natural and social systems within an application context of managed resources or infrastructure, utilizing climate and environmental observations, model outputs, socioeconomic data and decision models. The research should also incorporate elements of regional/sub-regional climate science and associated environmental processes, socioeconomic impacts, technological capabilities, management institutions and policies, and decision processes including evaluation.

Consistent with this CCSP objective, the EPA sponsored the six research projects presented in this book to further our understanding of the potential impacts of, and responses to, climate variability and change. The work was funded through the ORD Science to Achieve Results (STAR) program that funds research grants in numerous environmental science and engineering disciplines through a competitive solicitation process and independent peer review. The EPA Global Program dedicates a significant portion of its resources to extramural research grants to capitalize on expertise in the academic community that complements the EPA's laboratory research, as well as research conducted by other federal agencies.

The six projects focus on a range of potential impacts, including the effects of climate change on air quality, water quality, fisheries, urban infrastructure, public health and wildfires. They also focus on a variety of geographic scales, ranging from the regional level (for example the Southwest US and the Northern Plains) to the basin and watershed level (such as San Joaquin Basin in California's Central Valley and Mackinaw River watershed in Illinois), to the urban level (for example the New York and Boston metropolitan areas). The projects provide valuable insights about potential strategies for reducing the site-specific impacts of climate variability and change. Some of the strategies that would reduce risks posed by climate change or exploit opportunities may be viewed as 'no regrets,' because they make sense whether or not the effects of climate change are realized. For example, enhanced responses to urban heatwaves can save lives now, whether or not the frequency and intensity of heatwaves change as the climate changes. Similarly, new crop varieties that are heat- and drought-resistant may reduce crop losses during hot, dry summers today, whether or not the frequency and intensity of droughts change as the climate changes.

The insights provided by the six projects are valuable because adaptation has a cost. The scarce natural and financial resources used to adapt to climate change could be used for other productive activities. In the vernacular of economics, there are opportunity costs to using scarce resources for adaptation. These costs must be carefully weighed when considering the tradeoffs between alternative adaptation strategies, and between adapting to the change, reducing the cause of the change, and living with the residual impacts.

The research projects and assessments like those reported in this book support informed discussion of climate variability and change issues by decision makers, stakeholders and the general public. The assessments provide timely and useful information to decision makers so they can utilize the science to strengthen their environmental decisions, and implement programs and adaptation strategies that will increase the resilience of systems to climate variability and change. This type of decision support is an important mechanism for attaining the ultimate goal of meaningful improvements in human health and environmental quality.

REFERENCES

Intergovernmental Panel on Climate Change (IPCC) (2001), *Climate Change 2001: The Scientific Basis, Contribution of Working Group I to the Third Assessment Report of the Intergovernmental Panel on Climate Change*, J.T. Houghton, Y. Ding, D.J. Griggs, M. Noguer, P.J. van der Linden, X. Dai, K. Maskell and C.A. Johnson (eds), Cambridge: Cambridge University Press, p. 38.

Keeling, C.D. and T.P. Whorf (2005), 'Atmospheric CO_2 Records from Sites in the SIO Air Sampling Network', in *Trends: A Compendium of Data on*

Global Change, Oak Ridge: Carbon Dioxide Information Analysis Center, Oak Ridge National Laboratory, US Department of Energy.

National Academy of Sciences (1992), *Policy Implications of Greenhouse Warming: Mitigation, Adaptation, and the Science Base*, Washington, DC: National Academy Press.

National Research Council (2001), *The Science of Regional and Global Change: Putting Knowledge to Work*, Committee on Global Change Research, Washington, DC: National Academy Press.

Scheraga, J.D. and A.E. Grambsch (1998), 'Risks, Opportunities, and Adaptation to Climate Change', *Climate Research*, **10**: 85–95.

US Climate Change Science Program (CCSP) (2003), *Strategic Plan for the US Climate Change Science Program*, Washington, DC, July 2003.

1   Introduction

M. Ruth, K. Donaghy and P. Kirshen

LOCAL IMPACTS AND RESPONSES TO CLIMATE CHANGE AND VARIABILITY

This book presents results from six research projects on the impacts of and responses to climate change and climate variability. The projects have been carried out over the course of more than three years as part of the US Global Change Research Program (US GCRP). Six separate research groups concentrated on a diverse set of topics – from changes in wildland fire dynamics to water use in agriculture, to air quality impacts on human health, to reliability of urban infrastructure systems – all under a wide range of climate, socioeconomic and management scenarios.

The six projects present integrated assessments of the impacts of climate change, and adaptive and mitigating responses to it at urban and regional scales. These assessments have contributed to knowledge of localized experiences of climate change, how it affects different sectors, how different stakeholders perceive its implications and are adapting to it, and how decision support systems can serve to promote dialogues between researchers, stakeholders and policy makers.

A key feature that unites the chapters in this book is their emphasis on the development of approaches for integrated assessments of the potential consequences of climate variability and change on the United States. Specific emphasis is given to assessments that integrate observed and anticipated dynamics 'horizontally' and 'vertically'. Horizontal integration means analysis that brings together under a unified framework studies of impacts on, and responses by different sectors, such as water management, agricultural production and fisheries. Vertical integration means analysis from the level of climate systems through to socioeconomic impacts and responses.

Each of the six studies focuses on a finer geographic scale than is customary in integrated assessment research. Instead of broad global or continental-scale impacts, the research investigates consequences of climate change and variability at sub-national scales – such as individual river basins and mountain ranges or specific metropolitan areas. All six studies explore major trends of both human-induced and natural climate change and variability for the subsequent 25 to 100 years. Their purpose is to generate insights into climate change impacts that are best assessed at those fine geographic scales and that are of potentially significant environmental, social and/or economic importance. Specific interest lies in impacts which, when considered jointly, are likely to identify important interactions that would alter conclusions about the vulnerability of a locality or resource to climate change (US GCRP 2004). The result of this research is a rich set of insights into methodologies for integrated assessments and a set of guides for investment and policy making.

The findings from the projects have clear policy relevance – they are targeted at the levels at which impacts of climate change and climate variability are felt most acutely and at which the interests and capacities for change lie (Wilbanks and Kates 1999). In several instances, the research projects have made extensive use of insights of stakeholders to facilitate information exchange and make findings relevant to the groups and communities who are ultimately expected to act on the findings that the research results generate. Where appropriate, stakeholder participation was made an integral part of the research projects – from problem definition to data collection and model development, to interpretation of results and response strategies. In the process, stakeholder involvement advanced knowledge both among stakeholders and researchers. In several cases, the projects began to leverage existing capacities and provided avenues for the research results to find application in actual environmental investment and policy making in the light of climate change and climate variability.

The following section briefly reviews the causes and ramifications of climate change and variability. Next we discuss responses that may be chosen to reduce anthropogenic impacts on the climate and to prepare for continued climate change and variability, and the roles that stakeholders may play in assessment of climate impacts and identification of responses. Following this discussion of climate impacts and responses, we place the chapters of this book in the broader context of impact assessment research.

A PRIMER ON CLIMATE CHANGE AND VARIABILITY

Physical and Biogeochemical Processes

Earth's climate is regulated, in part, by the presence of gases and particles in the atmosphere which are penetrated by short-wave radiation from the sun and which trap the longer-wave radiation that is reflecting back from Earth.

Collectively, those gases are referred to as greenhouse gases (GHGs) because they can trap radiation on Earth analogous to the glass of a greenhouse and have a warming effect on the globe. The main GHG is water, which affects the overall energy budget of the globe and – working like a steam heating system – funnels energy through the hydrological cycle across regions. Among the other most notable GHGs are carbon dioxide (CO_2), methane (CH_4), nitrous oxide (N_2O) and chlorofluorocarbons (CFCs). Their sources include fossil fuel combustion, agriculture (e.g. releasing carbon from soils or methane from rice paddies and livestock) and industrial processes.

Each GHG has a different atmospheric concentration, mean residence time in the atmosphere, and different chemical and physical properties. As a consequence, each GHG has a different ability to upset the balance between incoming (solar) radiation and outgoing long-wave radiation. This ability to influence Earth's radiative budget is known as climate forcing. While some constituents of the atmosphere tend to reflect outgoing radiation back to Earth, the presence of aerosols in the atmosphere – released, for example, from coal burning power plants – leads to reflection of incoming radiation and thus has a cooling effect that may partly offset the warming effect of greenhouse gases (Wigley 1999).

Climate forcing varies across chemical species in the atmosphere. Spatial patterns of radiative forcing are relatively uniform for CO_2, CH_4, N_2O and CFCs because these gases are relatively long-lived and as a consequence become more evenly distributed in the atmosphere. In contrast, patterns of spatial radiative forcing of short-lived constituents, such as aerosols and ozone, are closely aligned with their sources of emissions (Wigley 1999).

Steep increases in atmospheric GHG and aerosol concentrations occurred since the Industrial Revolution. Those increases are unprecedented in Earth's history. As a result of higher GHG concentrations, global average surface temperature has increased by about 0.6°C during the 20th century with the 1990s as the warmest decade and 1998 the warmest year in the instrumental record since 1861 (IPCC 2001). These average global changes mask larger regional variations. For example, higher latitudes have warmed more than the equatorial regions (OSTP 1997).

A change in average temperatures may serve as a useful indicator of changes in climate, but it is only one of many ramifications of higher GHG concentrations. Since disruption of Earth's energy balance is neither seasonally nor geographically uniform; effects of climate disruption vary across space as well. For example, there has been a widespread retreat of mountain glaciers during the 20th century. Scientific evidence also suggests that there has been a 40 percent decrease in Arctic sea ice thickness during late summer to early autumn in recent decades and considerably slower decline in winter sea ice thickness, while Northern Hemisphere spring and summer ice extent have decreased by about 10–15 percent since the 1950s (IPCC 2001).

Large-scale efforts are under way to explore the complex causal relationships between human activities and climate change, to put the various pieces of the climate change puzzle together on computers, and to explore likely future climate conditions under alternative assumptions about biogeochemical mechanisms and human activities (IPCC 2001). A range of projections has emerged from these computer models, which indicate that global averaged surface temperature is likely to increase by 1.4–5.8°C over the period 1990–2100, making the projected rate of warming much larger than the observed changes during the 20th century and very likely larger than rates of warming for at least the last 10 000 years, as indicated by data derived from the paleoclimate record (Wigley 1999; IPCC 2001). Relative to any fixed threshold, the frequency of daily, seasonal and annual warm temperature extremes will likely increase and the frequency of daily, seasonal and annual cold weather extremes will likely decrease. As in the recent past, changes in temperature could be accompanied by larger year-to-year variations in precipitation, regionally distinct rates of snow and ice cover changes, and changes in sea level (Klein and Nicholls 1999; IPCC 2001).

Climate change models increasingly show climate responses that are consistent across very differently specified models, and responses that are consistent with recent observations. These models, combined with long-term historical analyses and field experiments, indicate that humanity has indeed embarked on a real-world climate change experiment of monumental proportions. Although increasingly sophisticated, the climate models on which predictions are based continue to suffer from uncertainties in many underlying biogeochemical processes and our fundamental inability to adequately anticipate future human responses to climate change. Moreover, the models' specifications make it difficult to reflect potential discontinuities of climate processes and instead often only portray gradual changes (Schelling 1992; Kay and Schneider 1994). Examples of discontinuities include rapid changes in the direction of ocean currents that funnel significant amounts of energy among continents and fundamentally affect regional temperature, sea levels and precipitation patterns. A gradual increase in temperature may result in local climate conditions that are unfavorable to some local species, triggering a change in species composition. Changes in species composition may affect diverse ecosystem features such as soil properties or pollination of fruit trees or crop species, local food supply and livelihoods, impact water regimes and spread of disease, and trigger changes in society and the economy.

Implications for Natural and Managed Ecosystems

Higher temperatures can lead to dramatic changes in the snowfall and snowmelt dynamics in mountainous watersheds and lead to more rapid, earlier and greater spring runoff (Gleick 1987; Jeton et al. 1996; Leung and Wigmosta 1999). This effect was already identified, for example, in the 1980s for watersheds in California (Burn 1994; Lettenmaier et al. 1994; Lins and Michaels 1994).

The net loss of snow and ice cover, combined with an increase in ocean temperatures and thermal expansion of the water mass in oceans, resulted in a rise of global average sea level between 0.1 and 0.2 meters during the 20th century, which is considerably higher than the average rate during the last several millennia (Barnett 1984; Douglas 2001; IPCC 2001). However, the rate and extent of sea level rise varies across the globe, with some areas losing heights relative to the sea, such as the UK and Western France, while others, such as Scandinavia and Scotland, are emerging (Doornkamp 1998). In some cases anthropogenic land subsidence – e.g. from mining, natural gas or ground water extraction – significantly speeds up the potential effects of climate change-induced relative rise of sea levels (Gampolati et al. 1999).

The impacts of changes in ocean temperatures, sea levels and coastal storm patterns are broad and include displacement and loss of wetlands, inundation of low-lying property, increased erosion of the shoreline, expansion of flood zones, and salinization of surface water and groundwater. Since many large cities and their built infrastructure are located on the coast, impacts from sea level rise on urban areas and their hinterlands will be significant.

Increasingly, public moneys will need to be directed towards protection of coastal aquifers and other public water supply systems (Nicholls et al. 1999). Locally, new infrastructure is put in place to protect coastal zones from inundation. Hard structures influence banks, channels, sediment deposits and morphology of the coastal zone, leading to a loss of coastal ecosystems (Sorenson et al. 1984; Weggel 1989; Gleick and Maurer 1990; Leatherman 1994). For example, estimates of the fixed costs for dikes or levees built to protect against a one-meter rise in sea level range from $150 to $800 per linear foot (1990 dollars) (ASCE 1992). Beach nourishment, one of the most popular soft protection strategies, may solve the fundamental problem of diminishing sediment resources, especially during the early onset of erosion (Yohe and Neumann 1997). However the long-term effectiveness of beach nourishment remains uncertain due to an incomplete understanding of coastal processes and their responses to future climate change (Neumann et al. 2000; Ruth and Kirshen 2001).

Changes in heat fluxes through the atmosphere and oceans, combined with changes in reflectivity of the earth's surface and an altered composition of GHGs and particulates in the atmosphere, may result in altered frequency and severity of climate extremes around the globe (Easterling et al. 2000; Meehl et al. 2000). For example, it is likely that there has been a 2–4 percent increase in the frequency of heavy precipitation events in the mid and high latitudes of the Northern Hemisphere over the latter half of the 20th century, while in some regions, such as Asia and Africa, the frequency and intensity of droughts have increased in recent decades (IPCC 2001). Furthermore, the timing and magnitude of snowfall and snowmelt may be significantly affected (Frederick and Gleick 2000), influencing, among others, erosion, water quality and agricultural productivity.

The proportion of total precipitation from heavy precipitation events has grown at the expense of moderate precipitation events (White and Howe 2002). As a result, flood magnitude and frequency tend to increase. Flooding is one of world's most costly and destructive natural disasters. It can seriously damage the built environment, paralyze transportation, interrupt energy distribution, impair wastewater plants, disrupt safe water supplies, pose threats to the health of species and humans, and even cause deaths or severe injury. For example, flooding in the UK during autumn 2000 caused an estimated £1 billion of damage and brought chaos to many parts of England and Wales (Zoleta-Nantes 2000). Floods in poor districts of Manila, Philippines, exposed people to respiratory infections, skin allergies and gastro-intestinal illnesses, with children most at risk (IPCC 2001)

The US, on average, is well-endowed with water. However fresh water can be a scarce resource virtually anywhere in the US at some time, especially in the urban areas of the arid and semiarid West (Alcamo et al. 2003). Despite the fact that detailed regional impacts of global climate change on future water supplies are notoriously uncertain (Frederick and Gleick 2000), consensus exists that climate change will affect the demand as well as the supply of water. It may substantially affect irrigation withdrawals (Doll and Siebert 2001). Net irrigation requirements per unit of irrigated area generally would decrease across much of the Middle East and northern Africa, whereas most irrigated areas in India and northern China would require more water (Boland 1997). The most sensitive areas in municipal water use to climate change are increased personal washing and increased use of water in gardens and for lawns (IPCC 2001). Industrial use for processing purposes is relatively insensitive to climate change. Demands for cooling water, in contrast, may be noticeably affected by climate change (Cruise et al. 1999; Frederick and Gleick 2000).

Climate change is also likely to affect water quality. Potential negative implications of climate change include lower flows, higher water temperatures, and increased storm surges. Lower flows in rivers will lead to increases in salinity levels to downstream water users and increase peak concentrations of metals and chemical compounds (IPCC 2001). Higher water temperatures alone would lead to increases in concentrations of some chemical species but decreases in others, and would also encourage the growth of algal blooms, which can lead to oxygen deficits in the water, and thus directly affect riverine ecosystems and indirectly the economies which depend on them (Frederick and Gleick 2000).

Increases in the number of days with more intense precipitation could increase the agricultural and urban pollutants washed into streams and lakes, further reducing oxygen levels (Frederick and Gleick 2000). However current understanding of the hydrological impacts is insufficient to determine whether climate change would improve or worsen low-flow conditions. The direction as well as the magnitude of the climate impacts on lake quality from changes in precipitation and evaporation rates is also uncertain (Frederick and Gleick 2000), and as a consequence the direct and indirect impacts on urban areas, as well as the needs for planning and investments are uncertain as well.

Impacts on the Health of Species

Sea level rise and even modest changes in the frequency, severity and distribution of tropical storms and hurricanes, for example, may have substantial impacts on coastal wetland patterns and processes, many of which are part of urban ecosystems. These impacts will combine with other human uses of wetlands, for example, agriculture, industry and settlements as well as the harvesting of plants and animals (Barth and Titus 1984; Carter 1988; Day et al. 1993; Michener et al. 1997). Fragmentation of landscapes, combined with changing climate conditions, may reduce diversity of indigenous species and prove an increasing challenge to the ability of natural resource managers to maintain viable habitats and species populations (Peters and Darling 1985; Peters and Lovejoy 1992). Disruptions of existing ecosystem processes may be the result, for example, of disruption of seed dispersal, limitations on foraging ranges, infringement on species migration corridors, or increased competition with exotic species (Fahrig and Paloheimo 1988).

Cities are often the ports of first entry of exotic species – introduced deliberately for agricultural production, ornamental uses or as pets, and inadvertently introduced in ballast waters of ships, or with agricultural and other products. The conditions for longer-term establishment of populations of exotic species may improve with climate change, in part because present ecosystems become increasingly stressed and because the conditions that favored their presence at their place of origin may now be found in their new destination. Since different species are likely to respond differently to climate change, changes in species composition may result (Graham 1988); and even though the extent to which disturbance in general affects invasion of ecosystems by species previously considered exotic (Lodge 1993), the abundance and diversity of exotic species is expected to increase (Sweeney et al. 1992).

Changes in precipitation and temperature regimes of urban areas also can lead to increased runoff of fertilizers and pesticides from intensively managed agricultural lands, parks and lawns; increased release of detergents and solvents from households and industry as a result of overwhelmed combined sewer overflow systems and water treatment plans; as well as runoff of oils and other petroleum products from roads, filling stations and parking lots during periods of heavy downpours. Hurricane-induced storm surges can have deleterious effects on inland freshwater and brackish wetlands and low-lying terrestrial areas because of the salt water, sediments and organic material that these surges carry inland (Blood et al. 1991; Knott and Matore 1991). Elevated salt levels may persist for more than a year, causing significant long-term changes in plant communities (Hook et al. 1991).

Climate asserts a significant influence on human health, as is evident by the geographic distribution and seasonal fluctuations of many diseases and causes of mortality (Tromp 1980). The connection between climate and human health strongly suggests that climatic change may alter the incidence and distribution of a wide range of diseases (Stone 1995) and mortality causes (Martens 1998). Public health researchers have, however, only recently begun to investigate the potential impacts of climate change and to identify adaptation strategies to reduce public health vulnerabilities to climate variability and change (Longstreth 1991; Kovats et al. 1999; Patz et al. 2000a; Patz et al. 2000b; WHO 2000; Watson and McMicheal 2001).

Changes in temperature and precipitation regimes, as well as in the frequency of extreme weather events will combine to affect morbidity and mortality. From a societal perspective, changes in extreme events may be an even larger concern than changes in climatic averages (Katz and Brown 1992; Changnon 2000). Recent research indicates that the frequency of extreme heat-stress events in the US may already have increased (Gaffen and Ross 1998). The recognition of likely future increases in extreme temperature events in combination with the well-established sensitivity of mortality to temperature extremes has resulted in expanded public health research to examine the effects of climate change on temperature-related mortality.

Mid-latitudinal climates exhibit strong cyclical temperature and mortality patterns (Lerchl 1998). Higher temperatures are commonly associated with lower mortality rates and, conversely, lower temperatures associated with higher mortality rates. The seasonal nature of mortality rates has been observed, for example, in heart failure-related morbidity and mortality (Steward et al. 2002), coronary heart disease (Pell and Cobbe 1999), and incidence of stroke (Lanska and Hoffmann 1999; Oberg et al. 2000). Some research indicates that the magnitude of the seasonal mortality oscillation may be dampening due to advances in medicine and the ability of humans to control their microenvironments (Seretakis et al. 1997; Lerchl 1998). Other researchers, however, find no evidence of a decline in the oscillation of seasonal mortality (Van Rossum et al. 2001).

Exposure to temperature extremes, such as those experienced during heatwaves and cold spells, is associated with rapid increases in mortality (Huynen et al. 2001). For example, more than 700 deaths in Chicago were attributed to the July 1995 heatwave (Semenza et al. 1996). Extreme heat events increase requirements on the cardiovascular system to produce physiological cooling which, in turn, may lead to excess deaths (Kilbourne 1997). In particular, infants, the elderly, individuals with pre-existing illnesses, the poor, the overweight and individuals living in urban areas are vulnerable to heat-related morbidity and mortality (Blum et al. 1998; Smoyer et al. 2000; CDC 2002; NWS 2002).

Extreme cold temperature events are also associated with increases in mortality rates, controlling for influenza (Kunst et al. 1993; Eurowinter Group 1997). Sharp increases in mortality during cold events have been identified, mainly due to thrombolic and respiratory disease (Donaldson and Keatinge 1997). Other mechanisms through which cold affects mortality include increases in blood pressure, blood viscosity and heart rate. Coronary and stroke mortality have been shown to be associated with cold temperatures in the US (Rogot and Padgett 1976). In Russia mortality is found to increase by 1.15 percent for each 1°C drop in temperature (Donaldson et al. 1998). A study of the impacts of temperature and snowfall on mortality in Pennsylvania found exposure to snow and temperatures below -7°C (19°F) to be dangerous to health (Gorjanc et al. 1999).

The effects of extreme temperature events on mortality are not solely determined by physiological variables, but also by the degree of acclimation of the local population to the regional climate regime (Kalkstein and Davis 1989; Kalkstein and Greene 1997; Smoyer 1998; Keatinge et al. 2000; Curriero et al. 2002). Acclimation entails the adaptation of communities to their environmental surroundings including behavioral patterns, societal fashions and customs such as dress and siestas, the thermal attributes of the local built infrastructure, availability of air conditioning and the health system's familiarity and ability to deal with weather-induced health conditions. In fact, research suggests that the sensitivity of mortality to extreme heat events has been decreasing over time, possibly as a result of societal adaptation (Davis et al. 2002).

The wide range of climatic environments inhabited by humans demonstrates our enormous ability to buffer ourselves from harsh macroenvironments. As an example, in Yakutsk, eastern Siberia – the coldest city in the world – no association is present between mortality rates and extremely cold temperatures (Donaldson et al. 1998). Yet, while acclimation enables a population to become less vulnerable to the prevalent weather events, the population remains susceptible to weather events that occur relatively infrequently (events at the tails of the probability distribution). Therefore, the changes in the frequency of extreme events accompanying climate change need to be examined in order to identify adaptation strategies such that the population can adapt to the characteristics of the new climate regime.

Studies investigating the impacts of climate on human health often employ a place-based approach in consideration of the importance of local acclimation in determining a population's morbidity and mortality (Martens 1998; Smoyer 1998). Place-specific mortality responses to changes in temperature have been found to be present even after controlling for differences in meteorological, demographic and economic variables (Smoyer et al. 2000). In general, mortality rates of populations in cool climates are more sensitive to heat events, whereas populations in warmer climates have mortality rates more sensitive to cold events (Curriero et al. 2002). To illustrate, Keatinge finds that for every 1°C decrease in temperature below 18°C (64.4°F) mortality rates in south Finland increases by only 0.27 percent while in Athens, Greece mortality rates increase by 2.15 percent (Keatinge 1997). Likewise, Kalkstein and Davis (1989) evaluate temperature-related mortality rates in 48 US cities and find considerable variation in heat threshold levels with, for example, heat thresholds in Phoenix and Las Vegas equal or exceeding 109°F (43°C) whereas in Boston and Pittsburgh the thresholds are below 86°F (30°C). A city-level study examining minimum mortality temperatures in 11 large US cities finds temperature differences of up to 15°F (8°C) between cities (Curriero et al. 2002).

Elevated temperatures not only result in heat stress – most notably among the elderly and urban poor – but also exacerbate local air pollution and thus air quality-related respiratory health problems. While one portion of society may increase their demand for air conditioning, potentially contributing to local energy shortages and urban heat island effects, others may increasingly suffer.

Though the intragenerational health effects of climate change may on occasion be notable, a host of non-climate related issues do play a major, if not overwhelming role in the health of a population (Smoyer et al. 2000). These non-climate related issues include changes in a population's age structure and ethnic diversity; economic prosperity; access to air conditioning, fresh water and health care; and integrity of social networks. Increased mobility too may lead to the spread of diseases irrespective of climate change.

Interrelationships Between Local and Global Climatic Conditions

Changes in land cover and land use affect local climatic conditions. For example, urban and other land use changes account for as much as half of the observed increases in the diurnal temperature range in the US (Kalnay and Cai 2003). Asphalt and concrete for roads, buildings and other structures necessary to accommodate growing populations absorb – rather than reflect – the sun's heat. The displacement of trees and shrubs eliminates the natural cooling effects of shading and evapotranspiration. Emission from energy conversion in power plants and combustion engines, especially when combined with reduced vegetation and larger areas with darker surfaces (Taha and Meier 1997) can raise air temperatures in a city by 2–8°F (1–3°C) (WMO 1984) and even change local temperature and precipitation patterns. The resultant 'heat island effect' is different from global warming, though it may exacerbate, and be exacerbated by climate variability and trends. Temperature increases and precipitation changes may stimulate further increases in energy use for cooling purposes, water pumping and more (US EPA 2000), and result in increased emissions of greenhouse gases, precursors of urban smog and contributors to changes in local environmental conditions.

Heat island effects have been observed most notably in urban areas. Changes in urban vegetation cover and albedo can be measured from space, using remote sensing, and correlated with climatologic information from urban weather stations. Time-series analyses, comparative time trends at one or more urban stations, comparisons along urban transects or among urban, suburban and rural stations, as well as between measurements on weekdays and weekends have helped document urban heat island effects for mega-cities across the US, including the New York Metropolitan area, Philadelphia, Washington DC, Pittsburg, Buffalo, Cleveland, Albany, Atlanta and Los Angeles (Bornstein and Lin 2000). Empirical evidence of urban heat island effects also exits for Turkey (Tayanc and Toros 1997), Austria (Böhm 1998), South Africa (Hughes and Balling 1996), Japan (Hadfield 2000), Singapore (Wong et al. 2003) and elsewhere. Although the name of urban heat island implies that it is solely an urban problem, research has shown urban heat islands are also becoming prevalent in small cities (Pinho and Orgaz 2000) and suburbs (Stone and Rodgers 2001).

Research into heat island effects suggests that heat island intensity decreases with increasing wind speed and increasing cloud cover (Ackerman 1985; Travis et al. 1987; Kidder and Essenwanger 1995; Figuerola and Mazzeo 1998; Magee et al. 1999; Morris et al. 2001; Unger et al. 2001). Heat island intensity most likely increases in the summer or warm half of the year (Schmidlin 1989; Klysik and Fortuniak 1999; Philandras et al. 1999) and tends to increase with increasing size of settlements and/or population (Park 1986; Yamashita et al. 1986; Hogan and Ferrick 1998; Torok et al. 2001). However several challenges to these generalizations have been mounted. For example, the greatest urban–rural difference detected in Birmingham, UK occurs in spring and autumn (Unwin 1980). Reykjavik, Iceland shows a tendency for negative heat island intensities (rural areas warmer than urban areas) in summer and only weak development at other times of the year (Steinecke 1999). A larger rate of growth of Prague's urban heat island has been detected since the 1920s in winter and spring than in summer (Brazdil and Budikova 1999).

Heat island phenomena affect the environment and population in a number of ways, including through increased demand for cooling energy, degradation of air quality, threats to public health, the triggering of adverse meteorological events and indirectly promoting urban sprawl. Increased energy demand for cooling and air conditioning are a direct result of higher ambient temperatures and decreased air quality. Increased energy demand, coupled with increasing energy prices, can result in greater costs to consumers. It is estimated that as much as 15 percent of the electricity consumed for cooling within Los Angeles is utilized for the sole purpose of offsetting the effects of the urban heat island (Rosenfeld and Romm 1996). The annual energy cost of urban heat islands alone within the US is estimated to be approximately $10 billion (Rosenfeld and Romm 1996).

Degradation of air quality, a result of increased emissions and higher ambient temperatures, may manifest itself in elevated concentrations of volatile organic compounds (VOCs), ground-level ozone and other air pollutants which may adversely affect the health of species, including humans (Cardelino and Chameides 1990). For example, ground-level ozone negatively impacts photosynthesis, inflames lung tissues and aggravates a range of respiratory ailments such as asthma. Researchers at the Lawrence Berkeley National Laboratory (LBNL) have estimated that each 1°F (0.6°C) rise in temperature over 70°F (21°C) increases the potential for ozone formation in Los Angeles by approximately 3 percent (US DOE 1996).

Heat islands may impact precipitation events either over or downwind of communities. Naturally occurring storms often intensify as they pass through cities. Moderate rainstorms may turn into full-blown thunder and lightning storms. The urban heat island in Atlanta, Georgia, for example, creates thunderstorms south of the city, which could cause urban flooding (NASA 1999). Urban heat islands have also been credited for torrential rains that wreaked havoc in Tokyo, Japan (Hadfield 2000).

Extreme temperature episodes, poor air quality and adverse meteorological conditions combine to worsen the habitability and comfort of human settlements in urban areas and may thus push people further away from those places. Yet, at the same time, complex and subtle relations among environmental conditions in urban and suburban areas may evade decision makers. For example, complex interactions of nitrogen oxides (NO_X) and urban ozone (O_3) may help reduce the potentially negative impacts of O_3 on plant growth in urban areas, while higher cumulative O_3 exposures and associated damages may result in suburban and rural areas with lower NO_X

concentrations (Gregg et al. 2003). As a result, urban heat island effects may make the problem of sprawl more intractable.

CLIMATE CHANGE AND RESPONSE STRATEGIES

Climate change may have many positive and negative, direct and indirect impacts on environmental, economic and social systems, and those impacts vary across space, time and various segments of an economy and society. Human settlement and resource use history is in large part characterized by adaptations to local environmental conditions. However, the scale and rate at which climate is changing poses new challenges for human response. Even if climate change impacts on socioeconomic systems are, by themselves, less than the combined non-climate impacts, their marginal effect could be significant, and they could noticeably compound existing stresses on resources, infrastructures and the institutions that govern their development and use.

To date the climate change debate has concentrated mainly on direct, negative impacts on current generations. Global response strategies have been identified to address what has been perceived, in essence, as a global problem. However, greater attention is being given recently to adaptation strategies, especially those that are beneficial even without climate change, and that lay the footprint for future development that is robust in the light of climate change. This section briefly reviews some of these strategies, concentrating on two broad categories – efforts to mitigate the greenhouse effect and measures to adapt to climate change. Both acknowledge that humans are not passive victims of climate change, and that simply insuring against adverse effects avoids the moral dimensions of climate change while jeopardizing the solvency of the insurance industry (Doornkamp 1998). The timing and extent of both mitigation and adaptation strategies are influenced by the tensions between the perceived needs, on the one hand, to resolve remaining uncertainties about climate change and, on the other hand, to be precautionary (Pearce 1991; Lemons and Brown 1995). The section closes with a discussion of the roles of stakeholders in climate impact assessment and identification of mitigation and adaptation strategies.

Mitigation

The United Nations Framework Convention on Climate Change (FCCC), which took effect in 1994, establishes as its goal the stabilization of GHG concentrations in the atmosphere 'at a level that would prevent dangerous anthropogenic interference with the climate system' (FCCC 1992). Towards that goal, parties to the convention are obliged to develop national inventories of GHG sources and sinks, to promote and cooperate in the development and diffusion of technologies that can prevent GHG emissions, to promote conservation and enhancement of GHG sinks and reservoirs, to cooperate in preparing for adaptation, to share information, and to promote education, training and public awareness. In addition, industrialized countries are asked to provide developing countries with financial resources to meet their commitments under the Framework Convention. In their 1997 annual meeting in Kyoto, the parties signed a protocol laying out mechanisms to achieve the Framework Convention's goals (FCCC 1997).

Common to the various mechanisms laid out in the Framework Convention is the intent to provide incentives to countries for reducing emissions beyond their own targets and to collaborate internationally to globally achieve cost-effective emissions reductions. Specific focus is given to economic incentives, such as marketable emissions permits, and to new institutions, such as the Global Environment Facility (GEF) to foster environmentally friendly development.

Promotion of technological change plays a crucial role in the climate change and development context (Edmonds et al. 2000). On the one hand, some changes in technology help boost output and reduce cost of fossil fuels, or energy end use devices. These changes tend to increase GHG emissions. On the other hand, efficiency improvements and increases in knowledge tend to decrease GHG emissions and cost of mitigation. The issue is further complicated by the fact that efficiency improvements and increases in knowledge are often related to production rates. Higher production and sales generate revenues for investment in new technology, and more experience is often gained as cumulative production increases (Yelle 1979; Ruth 1993). Furthermore, as relative prices of products change and development occurs, consumer preferences are likely to change. Substitution among inputs into production and among consumer goods and leisure activities – all of which are related to where people live – are key determinants of GHG emissions (Jorgenson et al. 2000). Yet little attention is paid in current international agreements to the indeterminacy of the net effects of technology change, technology transfer and changes in preferences for GHG emissions.

A slew of other instruments are already available in many countries to achieve specific emissions goals, or help leverage the effectiveness of market-based climate change policies. Among these instruments are environmental labeling requirements for electricity sources, demand side management, tax credits and accelerated depreciation schedules, planning and siting preferences for renewable energy facilities, renewable energy portfolio standards, land reclamation and reforestation policies, trace gas collection requirements for landfills, and more. Many of these instruments have originally been implemented to achieve goals such as improvements of energy security, achievements of higher ambient air quality, maintenance of ecosystem health and species diversity, or increased energy efficiency of households and firms. Coordination of their use may help further leverage GHG emission reductions (Dernbach 2000).

The Framework Convention's call for climate change mitigation has spurred a flurry of activities in government, industry and academia to identify for individual sectors of the economy how targets can be met and what the associated costs and benefits of alternative mitigation strategies are (Gwilliam 1993; Bernstein et al. 1999; Ruth et al. 2000). The debate quickly zeroed in on no-regrets strategies – strategies that are considered beneficial even if climate change were not an issue. Soon the debate proceeded to address how multiple policy instruments, ranging from taxes and subsidies to enhanced research and development efforts and regulatory instruments, could be applied simultaneously to more effectively improve efficiencies and reduce emissions (Ruth et al. 2000). More recently, the debate broadened to emphasize the wider range of social and environmental cost of energy use, aside from narrowly defined economic costs of energy conversion, GHG emissions and mitigation efforts (see for example Berry and Jaccard 2001). Solutions are being sought that transcend narrowly defined technological fixes and place technology policy in the broader context of development of adequate local capacity and essential support systems (Ruth et al. 2000). It is in this context that the relationships between urban development and climate change are being explored.

While social scientists and policy makers have begun to place climate change in the broader context of socioeconomic growth and development, natural scientists have begun to emphasize that non-CO_2 GHGs have caused most of the observed warming and that it may be more practical to reduce their emissions rather than emissions of CO_2, thus achieving climate goals more cost effectively (Hansen et al. 2000). How the confluence of these trends may be shaping climate change policy in the future is explored in more detail below, following a brief overview of the role of adaptation in dealing with climate change vulnerabilities.

Adaptation

Adaptation has often been perceived as the antidote to mitigation. Mitigation places emphasis on human capabilities to revert human-induced environmental trends. Adaptation, in contrast, means adjusting to climate change in order to reduce vulnerabilities of society and ecosystems, and is frequently perceived as an admission of an inability to noticeably revert climate change in a timely manner.

While successful mitigation depends on international cooperation, successful adaptation depends on local financial, technological and human resources. By the same token, mitigation has global benefits and adaptation has local benefits. As a consequence, mitigation has frequently been promoted as *the* proper response to the global issue of climate change. Yet, pursuit of adaptation strategies is neither an admission that climate change cannot be reverted, nor need it be a mere treatment of symptoms instead of eradication of the cause of the problem. As we discuss in more detail below, mitigation and adaptation can go hand in hand, and spending scarce resources on appropriate policy and investment strategies may successfully advance both mitigation and adaptation. Both also are closely related to land use, urban development and associated socioeconomic and technological issues.

Adaptation strategies can range from sharing or bearing losses, to actively reducing or preventing vulnerabilities. Some adaptations may occur as reactions to specific climate events, such as installations of pumps in basements and tunnels in response to increased rainfall, or increased chlorination of drinking water to prevent spread of diseases at higher temperatures. Others may be anticipatory, such as implementing early warning systems for extreme weather events, adjusting agriculture and forest management practices, genetically engineering crops, redesigning bridges to reduce scour at high-flow events, laying power lines underground to minimize susceptibility to wind and ice storms, or establishing habitat corridors for migratory species (Frankhauser 1996).

Much like some mitigation strategies, various adaptations to climate can generate benefits to society even if climate does not change. Benefits are derived from reducing susceptibilities to extreme weather events (Burton 1996) and correcting economic inefficiencies (Toman and Bierbaum 1996). Examples include changes in settlement patterns along rivers and coastlines that can help maintain healthy ecosystems that provide habitat for species, contribute to water retention and act as flood controls. In some instances retreat from susceptible areas may not be possible, making protection through biological barriers such as reforested mangrove forests, or artificial barriers such as sea walls, all the more relevant (Al-Farouq 1996).

The complex interrelations among climate, ecosystem health and socioeconomic development seem to call for a sophisticated set of strategies to address undesired outcomes. The fact that social and economic systems change rapidly with noticeable responses by the climate and ecosystems requires special focus on those geographic areas and sectors of an economy and society that are among the key drivers behind those changes. Consequently, the following section concentrates on urban systems.

The Roles of Stakeholders

In integrated assessments of the effects of climate change, the term 'science' applies to more than the knowledge of the workings of physical processes. It applies also to the interaction of social systems with ecosystems. As several of the studies discussed in this volume demonstrate, the impacts likely to be experienced in various locales will be determined by the adaptive and mitigative behavior of residents, policy makers and natural resource system managers – the many stakeholders in the public, private and non-profit realms.

If, as the US EPA program that funded these studies (STAR) intends, science is to achieve results, the science of integrated assessments of impacts of climate change at urban and regional scales – *as well as* the studies through which that science is developed – needs to be informed by stakeholders.

The type and extent of interactions between researchers and stakeholders and the extent to which the latter were integrated with the research projects over the period of time in which they were conducted varied from project to project, depending on research design. In some projects stakeholder input was critical to development of the research tools and decision support systems (DSSs).

Striking the right balance between stakeholder-involved science and stakeholder-informed science, without compromising the science itself, or using science to support agendas of select stakeholder groups, will be key to the future success of integrated assessments and decision support systems. This new kind of science will likely be guided by a high degree of social motivation, must meet the highest scientific standards, and will require a different organization, management and financial structure than is common in traditional environmental science. Several of the projects described in this book involved dozens of researchers and in some cases more than a hundred stakeholders – all with very different educational and professional backgrounds. All experienced long lead times to form effective research groups, faced severe budget constraints, and continued to run up against deadlines as new complexities in climate variability, climate change, impacts and response strategies were unveiled.

The managerial and leadership skills needed to insure project success is neither being taught to the next generation of scientists nor is it well-documented. Unrealized opportunities exist to build on the experiences laid out in this volume – and by similar projects around the world – to foster the dialog between science and society, and in the process of doing so to advance upon both.

OUTLINE OF THIS BOOK

The global climate change debate to date has heavily focused on anthropogenic emissions of greenhouse gases and the impacts of changing atmospheric concentrations of these gases on the stability of the climate system. Improved understanding of global climate change is used to point towards necessary mitigation strategies to avoid adverse feedbacks from climate variability and change to human living conditions and ecosystem processes. Goals for global reduction in greenhouse gas concentrations are being translated into international policies to guide national and regional development.

More recently, efforts have been increased to explore adaptation strategies that may reduce or avoid impacts of climate variability and change on local economies and ecosystems. Mechanisms to foster the implementation of adaptation strategies are being explored, and in the process of these recent developments, social science and planning-oriented analyses and modeling have expanded to complement biogeochemical models of climate change. While many of these developments are still driven by global concerns, the studies presented in this book focus heavily on local impacts and actions to improve quality of life through improvements in economic, social and ecosystem health.

The following chapter presents an assessment by Hugo Hidalgo and his co-workers of the impacts of climate change on water allocation, water quality and salmon production in the San Joaquin river basin, the most agriculturally intensive region in California's Central Valley. The study cuts across spatial scales, ranging from region-specific water runoff simulations, which are captured by soil moisture accounting models and snow accumulation and ablation models, to the large-scale drivers of climate change, reflected in the runs of General Circulation Models. A water allocation model is used to capture and simulate monthly impacts on irrigation, water storage and associated activities. The results point towards critical relationships between precipitation and temperature, on the one hand, and water availability and quality especially for smaller reservoirs, on the other hand.

Chapter 3 concentrates on interactions among wildland fires, climate change and variability, as well as societal dynamics in the Southwest US. Not only does the study in this chapter build on cutting-edge research into the dynamics of fires, the ecosystems within which those fires occur and the communities affected by fires, it also builds on the knowledge held by stakeholders in the region to inform wildland fire management. The integrated assessment combines stakeholder participation with geographic information systems and models to identify physical, biological and socioeconomic conditions at the scale of individual mountain ranges, and derives alternative management options for a variety of 'what-if' scenarios. Interface design, information presentation and communication of uncertainties in data, model structures and functional relationships are shown to be particularly critical for successful implementation of DSSs as means of conducting research and promoting interaction between researchers and stakeholders. The chapter points towards further research that is needed on the nature and quality of interaction between researchers and stakeholders both during the development and use of integrated assessments and DSSs to learn about, adapt to and mitigate the impacts of climate change.

Chapter 4 presents an integrated assessment of multiple-sector impacts on a Midwestern US watershed, produced by predicted changes in climate. The research used historical data, models, and standard and innovative analysis tools in conducting the assessment and was guided by early stakeholder input. The impact assessment focused on locations in the Mackinaw River watershed in Illinois. The project lays out sector specific responses to climate change; identifies relationships between and among sectors at each site, and among all sites; applies the impact analysis paradigm to identify and quantify local impacts produced by climate change; suggests mechanisms that produce an adaptive response to climate change while developing sector/system resilience to climate change impact; and integrates project results with a web-based decision support interface.

Chapter 5 presents an integrated assessment of the ecological and economic impacts of climate change on dryland grain production systems of the Northern Plains region of the US. The study explicitly captures adaptation actions as an endogenous driver, accounting for the fact that adaptation is typically not marginal but a clear deviation from past action. Special focus is given to the spatial and temporal variability in biophysical and economic conditions that result from different adaptation strategies.

In contrast to the preceding chapters, the remaining two studies of this book concentrate on urban systems. The research presented in Chapter 6, by Patrick Kinney and co-workers, focuses on potential public health impacts of climate change in the New York metropolitan area. This study presents an integration of global climate modeling with meso-scale meteorological modeling, land use/land cover change modeling and multi-scale air quality modeling to provide information about public health risks associated with changes in ambient air quality.

The Climate's Long-term Impacts on Metro Boston (CLIMB) project of Chapter 7 explores the potential impacts of climate change and variability on six major urban infrastructure systems and services: water supply and demand, water quality, flood control from riverine flooding and sea level rise, transportation, energy and public health. Analysis of each system and its services have been carried out under a wide range of climate, socioeconomic and technological scenarios, all of which are consistent across the sector-specific assessments and are integrated to explore adaptation options that help improve system performance. Like the research projects of the preceding chapters, the CLIMB project offers guidelines for environmental investment and policy strategies under a wide range of future scenarios. Like the companion studies in this book, the CLIMB project points towards future research and modeling needs to reduce uncertainties about climate impacts and responses. And, as in other chapters, it becomes clear that climate change has both positive and negative effects on the system under investigation. Some of the negative effects may be quite dramatic, others – especially when addressed by adequate adaptation strategies – may turn out to be rather benign.

Despite remaining uncertainties about future climate, socioeconomic and technological conditions, a set of responses can be identified to help reduce impacts of climate change and variability. Many of the strategies described in the following chapters are robust – are largely unaffected by the detailed assumptions about the climate, biophysical or socioeconomic conditions – and make good environmental and economic sense even in the absence of climate change. The concluding chapter returns to these lessons in more detail.

REFERENCES

Ackerman, B. (1985), 'Temporal march of the Chicago heat island', *Journal of Climate and Applied Meteorology*, **24**: 547–54.

Alcamo, J., M. Märker, M. Flörke and S. Vassolo (2003), 'Water and climate: a global perspective', *The Kassel World Water Series, Center for Environmental Systems Research*, University of Kassel.

Al-Farouq, A. (1996), 'Adaptation to climate change in the coastal resources sector of Bangladesh: some critical issues and problems', in J.B. Smith (ed.), *Adapting to Climate Change*, New York: Springer-Verlag, pp. 335–42.

American Society of Chemical Engineers (ASCE) (1992), 'Effects of sea-level rise on bays and estuaries', *Journal of Hydraulic Engineering*, **118**(1): 1–10.

Barnett, T.P. (1984), 'The estimation of global sea level change: a problem of uniqueness', *Journal of Geophysical Research*, **89**: 7980–88.

Barth, M. and J. Titus (1984), *Greenhouse Effect and Sea Level Rise*. New York: Van Nostrand Reinhold.

Bernstein, M., P. Bromley, J. Hagen and S. Hassell (1999), *Developing Countries and Global Climate Change: Electric Power Options for Growth*, Washington, DC: Pew Center on Global Climate Change.

Berry, T. and M. Jaccard (2001), 'The renewable portfolio standard: design considerations and an implementation survey', *Energy Policy*, **29**: 263–77.

Blood, E.R., P. Anderson, P.A. Smith, C. Nybro and K.A. Ginsberg (1991), 'Effects of Hurricane Hugo on coastal soil solution chemistry in South Carolina', *Biotropica*, **23**: 348–55.

Blum, L.N., L.B. Bresolin and M.A. Williams (1998), 'Heat-related illness during extreme emergencies', *Journal of the American Medical Association*, **279**(19): 1514.

Böhm, R. (1998), 'Urban bias in temperature time series - a case study for the city of Vienna, Austraia', *Climatic Change*, **38**(1): 113–28.

Boland, J.J. (1997), 'Assessing urban water use and the role of water conservation measures under climate uncertainty', *Climatic Change*, **37**: 157–76.

Bornstein, R. and Q. Lin (2000), 'Urban heat islands and summertime convective thunderstorms in Atlanta: three case studies', Atmospheric *Environment*, **34**(3): 507–16.

Brazdil, R. and M. Budikova (1999), 'An urban bias in air temperature fluctuations at the Klementinum, Prague, the Czech Republic', *Atmospheric Environment*, **33**: 4211–17.

Burn, D.H. (1994), 'Hydrologic effects of climatic change in West-central Canada', *Journal of Hydrology*, **160**: 53–70.

Burton, I. (1996), 'The growth of adaptation capacity: practice and policy', in J.B. Smith (ed.), *Adapting to Climate Change*, New York: Springer-Verlag: 55–67.

Cardelino, C.A. and W.L. Chameides (1990), 'Natural hydrocarbons, urbanization, and urban ozone', *Jounal of Geophysical Research*, **95**(D9): 13971–79.

Carter, R.W.G. (1988), *Coastal Environments*, New York: Academic Press.

Centers for Disease Control and Prevention (CDC) (2002), *Extreme Heat*, Atlanta, GA: Centers for Disease Control and Prevention.

Changnon, S. (2000), 'Human factors explain the increased losses from weather and climate extremes', *Bulletin of the American Meteorological Society*, **81**(3): 437–42.

Cruise, J.F., A.S. Limaye and N. Al-Abed (1999), 'Assessment of impacts of climate change on water quality in the southeastern United States', *Journal of the American Water Resources Association*, **35**: 1539–50.

Curriero, F.C., K.S. Heiner, K.S. Heiner, J.M. Samet, S.L. Zeger, L. Strug and J.A. Patz (2002), 'Temperature and Mortality in 11 cities of the eastern United States', *American Journal of Epidemiology*, **155**(1): 80–87.

Davis, R.E., P.C. Knappenberger, W.M. Novicoff and P.J. Michaels (2002), 'Decadal changes in heat-related human mortality in the eastern United States', *Climate Research*, **22**(2): 175–84.

Day, J.W., W.H. Conner, R. Costanza, G.P. Kemp and I.A. Mendelssohn (1993), 'Impacts of sea level rise on coastal systems with special emphasis on the Mississippi river deltaic plain', in R.A. Warrick, E.M. Barrow and T.M.L. Wigley (eds), *Climate and sea level change: observations, projections and implications*, Cambridge, England: Cambridge University Press.

Dernbach, J. (2000), 'Moving the climate change debate from models to proposed legislation: lessons from state experience', *Environmental Law Reporter*, **30**: 10933–79.

Doll, P. and S. Siebert (2001), *Global Modeling of Irrigation Water Requirements*, Kassel, Germany: University of Kassel.

Donaldson, G.C. and W.R. Keatinge (1997), 'Early increases in ischaemic heart disease mortality dissociated from, and later changes associated with, respiratory mortality, after cold weather in south east England', *Journal of Epidemiology and Community Health*, **51**(6): 643–48.

Donaldson, G.C., S.P. Ermakov, Y.M. Komarov, C.P. McDonald and W.R. Keatinge (1998), 'Cold related mortalities and protection against cold in Yakutsk, eastern Siberia: observation and interview study', *British Medical Journal*, **317**: 978–82.

Doornkamp, J.C. (1998), 'Coastal flooding, global warming and environmental management', *Journal of Environmental Management*, **52**: 327–33.

Douglas, B.C. (2001), 'An introduction to sea level', in B.C. Douglas, M.S. Kirney and S.P. Leatherman (eds), *Sea Level Rise: History and Consequences*, San Diego: Academic Press, pp. 1–11.

Easterling, D.R., G.A. Meehl, C. Parmesan and S.A. Changon (2000), 'Climate extremes: observations, modeling and impact', *Science*, **289**: 2068–74.

Edmonds, J., J.M. Roop and M.J. Scott (2000), *Technology and the Economics of Climate Change Policy*, Washington, DC: Pew Center on Global Climate Change.

Eurowinter Group (1997), 'Cold exposure and winter mortality from ischaemic heart disease, cerebrovascular disease, respiratory disease, and all causes, in warm and cold regions of Europe', *Lancet*, **349**: 1341–46.

Fahrig, L. and L. Paloheimo (1988), 'Effects of spatial arrangement of habitat patches on local population size', *Ecology*, **69**: 468–75.

Figuerola, P. and N. Mazzeo (1998), 'Urban-rural temperature differences in Buenos Aires', *International Journal of Climatology*, **18**: 1709–23.

Frankhauser, S. (1996), 'The potential cost of climate change adaptation', in J.B. Smith (ed.), *Adapting to Climate Change*, New York: Springer-Verlag, pp. 80–96.

Frederick, K.D. and P.H. Gleick (2000), *Water and Global Climate Change: Potential Impacts on US Water Resources*, Washington, DC: Pew Center on Global Climate Change.

Gaffen, D.J. and R.J. Ross (1998), 'Increased summertime heat stress in the US', *Nature*, **396**(6711): 529–30.

Gampolati, G., P. Teatini, L. Tomasi and M. Gonella (1999), 'Coastline regression of the Romagna region, Italy, due to natural and anthropogenic land subsidence and sea level rise', *Water Resources Research*, **35**(1): 163–84.

Gleick, P.H. (1987), 'Regional hydrologic consequences of increases in atmospheric carbon dioxide and other trace gases', *Climatic Change*, **10**(2): 137–61.

Gleick, P.H. and E.P. Maurer (1990), *Assessing the Costs of Adapting to Sea Level Rise: A Case Study of San Francisco Bay*, Oakland, CA: Pacific Institute for Studies in Development, Environment and Security and the Stockholm Environment Institute.

Gorjanc, M.L., W.D. Flanders, J. VanDerslice, J. Hersh and J. Malilay (1999), 'Effects of temperature and snowfall on mortality in Pennsylvania', *American Journal of Epidemiology*, **149**(12): 1152–60.

Graham, R.W. (1988), 'The role of climatic change in the design of biological reserves: the paleoecological perspective for conservation biology', *Conservation Biology*, **2**: 391–94.

Gregg, J.W., C.G. Jones and T.E. Dawson (2003), 'Urbanization effects on tree growth in the vicinity of New York City', *Nature*, **424**: 183–87.

Gwilliam, K.M. (1993), On Reducing Transport's Contribution to Global Warming, Paris: OECD.

Hadfield, P. (2000), 'Totally tropical Tokyo', *New Scientist*, **167**: 10.

Hansen, J., M. Sato, R. Ruedy, A. Lacis and V. Oinas (2000), 'Global warming in the twenty-first century: an alternative scenario', *Proceedings of the National Academy of Sciences*, **97**(18): 9875–80.

Hogan, A. and M. Ferrick (1998), 'Observations in nonurban heat islands', *Journal of Applied Meteorology*, **37**: 232–36.

Hook, D.D., M.A. Buford and M.A. Williams (1991), 'Impact of Hurricane Hugo on the South Carolina coastal plain forest', *Journal of Coastal Research*, **8**: 291–300.

Hughes W.S. and R.C. Balling (1996), 'Urban influences on South African urban temperature trends', *International Journal of Climatology*, **16**(8): 935–40.

Huynen, M., P. Martens, D. Schram, M. Weijenberg and A.E. Kunst (2001), 'The impacts of heatwaves and cold spells on mortality in the Dutch population', *Environmental Health Perspectives*, **109**: 463–70.

Intergovernmental Panel on Climate Change (IPCC) (2001), Climate Change 2001: Working Group II: Impacts, Adaptation and Vulnerability, Cambridge, UK.

Jeton, A.E., M.D. Dettinger and J. LaRue Smith (1996), Potential Effects of Climate Change on Streamflow of Eastern and Western Slopes of the Sierra Nevada, California and Nevada, Denver, CO: US Geological Survey, WRI Report, pp. 95–4260.

Jorgenson, D.W., R.J. Goettle, B.H. Hurt and J.B. Smith (2000), *The Role of Substitution in Understanding the Cost of Climate Change Policy*, Washington, DC: Pew Center on Global Climate Change.

Kalkstein, L.S. and R.E. Davis (1989), 'Weather and human mortality: an evaluation of demographic and interregional responses in the United States', *Annals of the Association of American Geographers*, **79**(1): 44–64.

Kalkstein, L.S. and J.S. Greene (1997), 'An evaluation of climate/mortality relationships in large US cities and the possible impacts of a climate change', *Environmental Health Perspectives*, **105**: 84–93.

Kalnay, E. and M. Cai (2003), 'Impact of urbanization and land use change on climate', *Nature*, **423**: 528–31.

Katz, R.W. and B.G. Brown (1992), 'Extreme events in a changing climate: variability is more important then averages', *Climatic Change*, **21**: 289–302.

Kay, J.J. and E.D. Schneide (1994), 'Embracing complexity', *Alternatives*, **20**(3): 32–39.

Keatinge, W.R. (1997), 'Cold exposure and winter mortality from ischaemic heart disease, cerebrovascular disease, respiratory disease, and all causes in warm and cold regions of Europe', *Lancet*, **349**(9062): 1341–46.

Keatinge, W.R., G.C. Donaldson, E. Cordioli, M. Martinelli, A.E. Kunst, J.P. Mackenbach, S. Nayha and I. Vuori (2000), 'Heat Related Mortality in Warm and Cold Regions of Europe: Observational Study', *British Medical Journal*, **321**(7262): 670.

Kidder, S. and O. Essenwanger (1995), 'The effect of clouds and wind on the difference in nocturnal cooling rates between urban and rural areas', *Journal of Applied Meteorology*, **34**: 2440–48.

Kilbourne, E.M. (1997), 'Heatwaves and hot environments', in E. K. Noji (ed.), *The Public Health Consequences of Disasters*, Oxford: Oxford University Press.

Klein, R.J.T. and R.J. Nicholls (1999), 'Assessment of coastal vulnerability to climate change', *Ambio*, **28**(2): 182–83.

Klysik, K. and K. Fortuniak (1999), 'Temporal and spatial characteristics of the urban heat island of Lodz, Poland', *Atmospheric Environment*, **33**: 3885–95.

Knott, D.M. and R.M. Matore (1991), 'The short-term effects of Hurricane Hugo on fishes and decapod crustaceans in the Ashley River and adjacent marsh creeks, South Carolina', *Journal of Coastal Research*, **8**: 335–56.

Kovats, R.S., A. Haines, R. Stanwell-Smith, P. Martens, B. Menne and R. Bertollini (1999), 'Climate change and human health in Europe', *British Medical Journal*, **318**: 1682–85.

Kunst, A.E., C.W. Looman and J.P. Mackenbach (1993), 'Outdoor air temperature and mortality in The Netherlands: a time-series analysis', *American Journal of Epidemiology*, **137**(3): 331–41.

Lanska, D.J. and R.G. Hoffmann (1999), 'Seasonal variation in stroke mortality rates', *Neurology*, **52**: 984–90.

Leatherman, S. (1994), *Coastal Resource Impacts and Adaptation Assessment Methods*, College Park, MD: University of Maryland.

Lemons, J. and D.A. Brown (1995), *Sustainable development: science, ethics and public policy*, Dortrecht: Kluwer Academic.

Lerchl, A. (1998), 'Changes in the seasonality of mortality in Germany form 1946 to 1995: the role of temperature', *International Journal of Biometeorology*, **42**: 84–88.

Lettenmaier, D.P., E.F. Wood and J.R. Wallis (1994), 'Hydro-climatological trends in the continental United States 1948–1988', *Journal of Climate*, **7**: 586–607.

Leung, R. and M.S. Wigmosta (1999), 'Potential climate change impacts on mountain watersheds in the Pacific Northwest', *Journal of the American Water Resources Association*, **35**(6): 1463–71.

Lins, H.F. and P.J. Michaels (1994), 'Increasing US streamflow linked to greenhouse forcing', *EOS, Transactions, American Geophysical Union*, **75**(281): 284–85.

Lodge, D.M. (1993), 'Species invasions and deletions: community effects and responses to climate and habitat change', in P.M. Kareiva, J.G. Kingsolver and B. Huey (eds), *Biotic Interactions and Global Change*, Sunderland, MA: Sinauer Associates, pp. 367–87.

Longstreth, J. (1991), 'Anticipated public health consequences of global climate change', *Environmental Health Perspectives*, **96**: 139–44.

Magee, N., J. Curtis and G. Wendler (1999), 'The urban heat island effect at Fairbanks, Alaska', *Theoretical and Applied Climatology*, **64**(1-2): 39–47.

Martens, W.J.M. (1998), 'Climate change, thermal stress and mortality changes', *Social Science and Medicine*, **46**(3): 331–44.

Meehl, G.A., T. Karl, D.R. Easterling and S.A. Changnon (2000), 'An introduction to tends in extreme weather and climate events: observations, socioeconomic impacts, terrestrial ecological impacts, and model projections', *Bulletin of the American Meteorological Society*, **81**(3): 413–16.

Michener, W.K., E.R. Blood, K.L. Bildstein, M.M. Brinson and L.R. Gardner (1997), 'Climate change, hurricanes and tropical storms, and rising sea level in coastal wetlands', *Ecological Applications*, **7**: 770–801.

Morris, C., I. Simmonds and N. Plummer (2001), 'Quantification of the influences of wind and cloud on the nocturnal urban heat island of a large city', *Journal of Applied Meteorology*, **40**: 169–82.

National Aeronautics and Space Administration (NASA) (1999), 'Atlanta's urban heat island', Washington, DC: National Aeronautics and Space Administration.

National Weather Service (NWS) (2002), *Heatwave*, Washington, DC: National Weather Service., NOAA, http://www.noaa.gov, last accessed 13 June 2004.

Neumann, J.E., G. Yohe, R. Nicholls and M. Manion (2000), *Sea Level Rise and Global Climate Change: A Review of Impacts to US Coasts*, Washington, DC: Pew Center on Global Climate Change.

Nicholls, R.J., F.M.J. Hoozemans and M. Marchand (1999), 'Increasing flood risk and wetland losses due to global sea-level rise: regional and global analyses and the science of climate change', *Global Environmental Change*, **9**: 569–87.

Oberg, A.L., J.A. Ferguson, L.M. McIntyre and R.D. Horner (2000), 'Incidence of stroke and season of the year: evidence of an association', *American Journal of Epidemiology*, **152**(6): 558–64.

Office of Science and Technology Policy (OSTP) (1997), *Climate Change: State Of Knowledge*, Washington, DC: Office of Science and Technology Policy, Executive Office of the President.

Park, H. (1986), 'Features of the heat island in Seoul and its surrounding cities', *Atmospheric Environment*, **20**: 1859–66.

Patz, J.A., D. Engelberg and J. Last (2000a), 'The effects of changing weather on public health', *Annual Review of Public Health*, **21**: 271–307.

Patz, J.A., M. McGeehin, S.M. Bernard, K.L. Ebi, P.R. Epstein, A. Grambsch, D.J. Gubler, P. Reiter, I. Romieu, J.B. Rose, J.M. Samet and J. Trtanj (2000b), 'The potential health impacts of climate variability and change for the United States: Executive Summary of the Report of the Health Sector of the US National Assessment', *Environmental Health Perspectives*, **108**: 367–76.

Pearce, D. (1991), 'Evaluating the socioeconomic impacts of climate change: an introduction', in D. Pearce (ed.), *Climate Change: Evaluating the Socioeconomic Impacts*, Paris: OECD, pp. 9–20.

Pell, J.P. and S.M. Cobbe (1999), 'Seasonal variations in coronary heart disease', *Quarterly Journal of Medicine*, **92**: 689–96.

Peters, R.L. and J.D. Darling (1985), 'The greenhouse effect and nature reserves', *BioScience*, **35**: 707–17.

Peters, R.L. and T.E. Lovejoy (1992), *Global Warming and Biological Diversity*, New Haven, CT: Yale University Press.

Philandras, C., D. Metaxas and P.T. Nastos (1999), 'Climate variability and urbanization in Athens', *Theoretical and Applied Climatology*, **63**: 65–66.

Pinho, O. and M. Orgaz (2000), 'The urban heat island in a small city in coastal Portugal', *International Journal of Biometeorology*, **44**(4): 198–203.

Rogot, E. and S.J. Padgett (1976), 'Associations of coronary and stroke mortality in with temperature and snowfall in selected areas of the United States, 1962–1966', *American Journal of Epidemiology*, **103**(6): 565–75.

Rosenfeld, A.H. and J.J. Romm (1996), 'Policies to reduce heat islands: magnitudes of benefits and incentives to achieve them', Proceedings of the 1996 LBL-38679 ACEEE Summer Study on Energy Efficiency in Buildings, Pacific Grove, CA.

Ruth, M. (1993), Integrating Economics, Ecology and Thermodynamics, Dortrecht: Kluwer Academic.

Ruth, M. and P. Kirshen (2001), 'Integrated impacts of climate change upon infrastructure systems and services in the Boston Metropolitan Area', *World Resource Review*, **13**(1): 106–22.

Ruth, M., A. Amato and B. Davidsdottir (2000), 'Impacts of market-based climate change policy on the US iron and steel industry', *Energy Sources*, **22**(3): 269–80.

Schelling, T.C. (1992), 'Some economic of global warming', *The American Economic Review*, **82**(1): 1–14.

Schmidlin, T. (1989), 'The urban heat island at Toledo, Ohio', *Ohio Journal of Science*, **89**: 38.

Semenza, J.C., C.H. Rubin, K.H. Falter, J.D. Selanikio, W.D. Flanders, H.L. Howe and J.L. Wilhelm (1996), 'Heat-related deaths during the July 1995 heatwave in Chicago', *The New England Journal of Medicine*, **335**(2): 84–90.

Seretakis, D., P. Lagiou, L. Lipworth, L.B. Signorello, K.J. Rothman and D. Trichopoulos (1997), 'Changing seasonality of mortality from coronary heart disease', *Journal of the American Medical Association*, **278**(12): 1012–14.

Smoyer, K.E. (1998), 'Putting risk in its place: methodological considerations for investigating extreme event health risks', *Social Science and Medicine*, **47**(11): 1809–924.

Smoyer, K.E., D.G.C. Rainham and J.N. Hewko (2000), 'Heat-stress-related mortality in five cities in Southern Ontario: 1980–1996', *International Journal of Biometeorology*, **44**: 190–97.

Sorensfon, R.M. and R.N. Weisman (1984), 'Control of erosion, inundation and salinity intrusion caused by sea level rise', in M.C. Barth and J.G. Titus (eds), *Greenhouse Effect and Sea Level Rise*, New York: Van Nostrand Reinhold, pp. 179–214.

Steinecke, K. (1999), 'Urban climatological studies in the Reykjavik subarctic environment, Iceland', *Atmospheric Environment*, **33**: 4157–62.

Steward, S. and M.B. McIntyre (2002), 'Heart failure in a cold climate: seasonal variation in heart failure-related morbidity and mortality', *Journal of the American College of Cardiology*, **39**(5): 760–66.

Stone, B.J. and M.O. Rodgers (2001), 'Urban form and thermal efficiency: how the design of cities influences the urban heat island effect', *Jounal of the American Planning Association*, **67**(2): 186–98.

Stone, R. (1995), 'If the mercury soars, so may health hazards', *Science*, **267**(5200): 957–58.

Sweeney, B.J., J.K. Jackson, J.D. Newbold and D.H. Funk (1992), 'Climate change and the life histories and biogeography of aquatic insects in eastern North America', in P. Firth and S.G. Fisher (eds), *Global Climate Change and Freshwater Ecosystems*. New York: Springer-Verlag. pp. 143–76.

Taha, H. and A. Meier (1997*), Mitigation of Urban Heat Islands: Meteorology, Energy, and Air Quality Impacts*, Proceedings of the International Symposium on Monitoring and Management of Urban Heat Island, Keio University, Fujisawa, Japan.

Tayanc, M. and H. Toros (1997), 'Urbanization effects on regional climate change in the case of four large cities in Turkey', *Climatic Change*, **35**: 501–24.

Toman, M. and R. Bierbaum (1996), 'An overview of adaptation to climate change', in J.B. Smith (ed.), *Adapting to Climate Change*, New York: Springer-Verlag, pp. 5–15.

Torok, S.J., C.J.G. Morris, C. Skinner and N. Plummer (2001), 'Urban heat island features of southeast Australian towns', Australian Meteorological Magazine, 50(1): 1–13.

Travis, D., V. Meentemeyer and P.W. Suckling (1987), 'Influence of meteorological conditions on urban/rural temperature and humidity differences for a small city', *Southeastern Geographer*, **27**: 90–100.

Tromp, S.W. (1980), Biometeorology: The Impact of the Weather and Climate on Humans and Thier Environment (Animals and Plants), Philadelphia, PA: Heyden and Sons.

Unger, J., Z. Sümeghy and J. Zoboki (2001), 'Temperature cross section features in an urban area', *Atmospheric Research*, **58**(2): 117–27.

United Nations Framework Convention on Climate Change (FCCC) (1992), United Nations Framework Convention on Climate Change, New York, United Nations.

United Nations Framework Convention on Climate Change (FCCC) (1997), Kyoto Protocol to the United Nations Framework Convention on Climate Change, New York, United Nations.

Unwin, D. (1980), 'The synoptic climatology of Birmingham's heat island', *Weather*, **35**: 43–50.

US Department of Energy (US DOE) (1996), Working to cool urban heat islands, Berkeley National Laboratory PUB-775, Berkeley, CA.

US Environmental Protection Agency (US EPA) (2000), 'Heat island effects', Washington, DC: US Environmental Protection Agency.

US Global Change Research Program (US GCRP) (2004), *US Global Change Research Program*, http://www.usgcrp.gov/, last accessed 13 September 2004.

Van Rossum, C.T.M., M.J. Shipley, H. Hemingway, D.E. Grobbee, J.P. Mackenbach and M.G. Marmot (2001), 'Seasonal variation in cause-specific mortality: are there high-risk groups? 25-year follow-up of civil servants from the first Whitehall study', International Journal of Epidemiology, 30: 1109–16.

Watson, R.T. and A.J. McMicheal (2001), 'Global Climate Change – the latest assessment: does global warming warrant a health warning?', *Global Change and Human Health*, **2**(1): 64–75.

Weggel, J.R. (1989), 'The cost of defending developed shoreline along sheltered shores', in J.B. Smith and D. Tirpak (eds), *Potential Effects of Global Climate Change on the United States*, Washington, DC: US Environmental Protection Agency.

White, I. and J. Howe (2002), 'Flooding and the role of planning in England and Wales: a critical review', *Journal of Environmental Planning and Management*, **45**(5): 735.

Wigley, T.M.L. (1999), *The Science of Climate Change*. Washington, DC, Pew Center on Global Climate Change.

Wilbanks, T.J. and R.W. Kates (1999), 'Global change in local places: how scale matters', *Climatic Change*, **43**: 601–28.

Wong, N., S. Tay, R. Wong, C.L. Ong and A. Sia (2003), 'Life cycle cost analysis of rooftop gardens in Singapore', *Building and Environment*, **38**(3): 499–509.

World Health Organization (WHO) (2000), *Climate Change and Human Health: Impact and Adaptation*. Geneva: World Health Organization.

World Meteorological Organization (WMO) (1984), 'Urban Climatology and its Applications with Special Regard to Tropical Areas', *Proceedings of the Technical Conference Organized by the World Meteorological Organization*, Mexico.

Yamashita, S., K. Sekine, M. Shoda, K. Yamashita and Y. Hara (1986), 'On relationships between heat island and sky view factor in the cities of Tama River basin, Japan', *Atmospheric Environment*, **20**: 681–86.

Yelle, L.E. (1979), 'The learning curve: historical survey and comprehensive survey', *Decision Sciences*, **10**: 302–34.

Yohe, G. and J. Neumann. (1997), 'Planning for sea level rise and shore protection under climate uncertainty', *Climatic Change*, **37**: 111–40.

Zoleta-Nantes, D.B. (2000), 'Flood hazard vulnerabilities and coping strategies of residents of urban poor settlements in Metro Manila, The Philippines', in D.J. Parker (ed.), *Floods*, London: Routledge, pp. 69–78.

2 Assessment of the Impacts of Climate Change on the Water Allocation, Water Quality and Salmon Production in the San Joaquin River Basin

H. Hidalgo, L. Brekke, N. Miller, N. Quinn, J. Keyantash and J. Dracup

INTRODUCTION

California's climate and geography make its agro-economic and environmental welfare particularly vulnerable to the impacts of climate change. The majority of California's water supply (without the Colorado River imports) arrives during the winter months in the high elevations. This water is stored as snowpack and is available only during the spring snowmelt. Conversely, a large part of the water demand for agricultural uses occurs during the summer months, particularly in the Central Valley. This disjoint timing of water supply relative to demand imposes challenges in the management of California's water resources (Hidalgo et al. 2005).

Climate change could potentially alter different aspects of the water supply in the headwaters and of the demand in the Central Valley. These changes will ultimately translate into impacts in the environment and agricultural production, an essential source of income and employment for the state. California is the highest ranked state in terms of added-value of agricultural production and the sixth largest agricultural exporter in the world (DWR 1998). The different agricultural activities in the Central Valley represent approximately 6–7 percent of the total income, employment and added-value of the state's trillion-dollar economy. In the San Joaquin Valley, the most agriculturally intensive region in the Central Valley, agriculture represents 32 percent of the total income, 28 percent of the added-value and 37 percent of the employment of the region (1998 agro-statistics from Kuminoff et al. 2000). Competing with these interests, there is an increasing awareness of protecting the environment of the Valley and the San Francisco Bay Delta and estuary. This awareness has been translated into a complex set of water quantity and quality regulations throughout the main agricultural producing areas and headwaters (see for example 1992 Central Valley Project Improvement Act, 1995 Delta Water Quality Control Plan, 2000 CALFED Record of Decision). The regulations manifest in the form of operational constraints on reservoir levels, exports of water from the Sacramento Valley to South-of-Delta regions, and salinity levels at certain points and times in the Delta and San Joaquin River (SJR) network.

This chapter contains the main results from an assessment on the potential impacts for water resources in the San Joaquin River Basin (SJRB) of California funded by the US Environmental Protection Agency (US EPA) through the Science to Achieve Results (STAR) program (Grant R827448). This assessment contributes to the collection of studies that characterize potential climate change impacts on Central Valley water resources (for example Gleick 1987; Lettenmaier and Gan 1990; US BR 1991; Dracup and Pelmulder 1993; US EPA 1997; Knowles and Cayan 2002; Miller et al. 1999, 2003). A summary of the findings of these previous studies is presented in Miller et al. (2003).

Consistent with the objectives of the STAR program, a chief objective in this project was to develop sufficient guidelines that can help stakeholders direct their capital improvement efforts and water resource planning in a way that accommodates potential climate change impacts. Given the fiscal requirements necessary for implementing capital improvements, it is important for resource managers and policy makers to understand the uncertainty of potential climate change impacts before investing in mitigation projects and adaptation programs. In particular, one of the main objectives of this research is to determine the potential range of climate change impacts on the SJRB's water resources, including reservoir inflows, reservoir operations (such as storage, releases and deliveries), and total system operations to meet water quality objectives of the Delta. Water quality aspects currently receive a high priority in terms of SJR water management objectives. The success of these efforts will ultimately depend on the ability to manage limited regional water resources. Climate change holds potential for disrupting this resource balance.

There are numerous ecological effects determined by altered streamflow quantity and/or timing that could have been considered in this assessment. In this study an example will be presented to focus upon one important response: SJRB populations of Chinook salmon, *Oncorhynchus tschawytscha*, linked to the changes in the abundance of cold freshwater. Consequently, our approach was to estimate future Chinook salmon populations due to hydrological changes in the SJRB. This task was accomplished by constructing an artificial neural network (ANN) model and the results of which are presented towards the end of this chapter.

The approach on the present climate change assessment involves using an assumption of the future global rate of CO_2 increase, simulating the global climate response to this rate of CO_2 increase, statistically downscaling global climate response to regional climate response, and translating regional climate change into impacts on streamflow, water allocation, water quality and the environment at the regional scale. A parallel approach was implemented where the starting point was simply a regional climate change assumption for key hydrologic basins in California's Central Valley. Regional assumptions were made on potential average annual changes to precipitation (P) and temperature (T). From this point on, the regional-based assessment methodology was the same as with that based on global climate change assumptions. Hydrologic and allocation/operations impacts from both sets of assessments are discussed in subsequent sections on climate projections and water allocation simulations. The results from this study will produce climate change impacts information on key water-related variables in the SJRB and greater Central Valley. This information has practical applications for water managers and SJRB stakeholders.

THE SAN JOAQUIN RIVER BASIN MANAGED SYSTEM

Water system management in the SJRB (Figure 2.1) was chosen as a relevant case study for assessing California's vulnerabilities to climate change because the region is highly productive agriculturally while having limited water supplies. The water systems that make this fact possible feature an elaborate array of surface water storage and conveyance systems. The functionality of these systems varies for irrigators on the West or East Side of the region. East Side irrigators are supplied by Sierra Nevada runoff via local water systems along each of the SJR's main tributaries: the Stanislaus River, the Tuolumne River, and the Merced River. West side irrigators receive supplies from the Sacramento River region via the federal Central Valley Project (CVP). The CVP includes in-stream reservoirs in the Sacramento River region (which includes Lake Shasta, Trinity Reservoir and Folsom Lake) and conveyance facilities to deliver these supplies to contractors in the SJRB region. Critical CVP conveyance facilities include pumping facilities in the San Francisco Bay-Delta (such as the Tracy Pumping Facility), which divert converged runoff from the Sacramento and SJRB regions through the Delta-Mendota Canal (DMC). Water pumped and sent south in the DMC serves agricultural demands or is stored in the off-stream San Luis Reservoir (SLR). The State of California operates a parallel water project that has similar functionality as the

CVP. The state system, referred to as the State Water Project (SWP), contains North-of-Delta storage, Delta export pumps (such as the Banks Pumping Facility), and South-of-Delta conveyance facilities to serve a community of state water contractors primarily located in Southern California. The coordinated Delta operations of CVP and SWP systems play a large role in determining the reliability of surface water deliveries for West side irrigators in the SJRB region and Southern California SWP contractors.

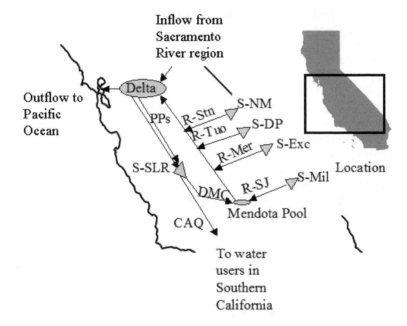

Notes: The main stem of the San Joaquin river has four main tributaries, the San Joaquin river (R-SJ), the Merced river (R-Mer), the Tuolumne river (R-Tuo) and the Stanislaus river (R-Stn). Four reservoirs are located at the respective headwaters of these rivers: Millerton Lake Reservoir (S-Mil), Exchequer Reservoir (S-Exc), Don Pedro Reservoir (S-DP) and New Melones Reservoir (S-NM). Delta water export facilities include the Delta Export (Pumping) Facilities (PPs), the federal export through the Delta Mendota Canal (DMC) and the State export through the California Aqueduct (CAQ). The confluence of the Sacramento and San Joaquin rivers occurs at the Delta, a major water quality management region.

Source: Adapted from Brekke et al. 2004

Figure 2.1   Location of the main tributaries and water-management facilities of the San Joaquin River Basin Region.

The increased salinity from agricultural production in the SJRB is one of the most concerning aspects of the degradation of water quality in the basin. The integrated contribution of salts from the SJRB to the Delta is regulated at the outlet of the system, in a point in the river known as Vernalis. Vernalis is the furthest downstream point in the SJR without tidal influence. The State Water Regional Control Board (SWRCB) has established a water quality objective for electrical conductivity (EC), an indicator of dissolved salts, at Vernalis of 700 μs cm^{-1} from April through August and 1000 μs cm^{-1} from September through March. Violation of the objective triggers water quality releases from New Melones Reservoir to dilute San Joaquin River water to meet the 30-day running average EC objective.

Another common water quality diagnostic variable representing the water discharge of the Central Valley to the Delta is the location of the X2 point. The X2 location is one of the water management standards adopted for Delta management and represents the minimum, tidally-averaged location (km) of the two 'practical salinity units' isohaline measured upstream from the Golden Gate Bridge. When the X2 is in the Suisun Bay (X2 is small), many of the estuary's living resources benefit (Brown and Kimmerer 2001). The basic assumption is that Suisun Bay presents a more productive habitat for some fish species. In contrast, when the X2 is located in the Delta (X2 is large), fish and other organisms are more vulnerable to be drawn into water project diversions (Brown and Kimmerer 2001). A reduction of the discharge of fresh water from the San Joaquin and Sacramento basins will increase the location of the X2, bringing the X2 point closer to the Delta and presumably endangering aquatic life.

CLIMATE PROJECTIONS FOR THE 21ST CENTURY

Hydrologic Impacts Based on Global Climate Change Assumptions

Using the mean of the Intergovernmental Panel on Climate Change (IPCC) Global Circulation Model (GCM) projections as a benchmark, a warm and wet GCM climate projection based on the Hadley Center's HadCM2 run one (hereafter HadCM) and a cool and dry climate projection based on the National Center for Atmospheric Research Parallel Climate Model (NCAR PCM) run B06.06 (hereafter PCM) were selected as end member climate scenarios for this assessment (Miller et al. 2003). These projections are based on a CO_2 emission scenario that considers a 1 percent per-year increase in the atmospheric mean global 'equivalent CO_2' relative to present-day conditions. The 'equivalent' rate represents increases in CO_2, as well as other greenhouse gases such as methane and NO_x (IPCC 2001). A 1 percent per-year increase in equivalent CO_2 is within the range of increases represented in the IPCC (2001) Special Report on Emissions Scenarios. From these coupled atmosphere–ocean GCM simulations, two 30-year periods (2010 to 2039, 2050 to 2079), and one 20-year period (2080 to 2099) were used.

GCM estimates of P and T were statistically downscaled to 10 km spatial resolution and integrated into monthly values using the PRISM (Precipitation-elevation Regression on Independent Slopes Model) technique (Daly et al. 1999). Monthly T shifts and P ratios were computed through comparison of GCM-simulated T and P with and without increases in CO_2 emissions. In general, PCM results suggest warmer and slightly drier conditions (compared to historical conditions) for California, while the HadCM estimates are associated with much warmer and much wetter conditions (Figure 2.2). The monthly increments were imposed on historical 1963 to 1992 T and P time-series for California to drive hydrologic models on six key California basins with natural flow data available for calibration. Hydrologic modeling in this study was performed using basin-specific applications of the National Weather Service River Forecast System Sacramento Soil Moisture Accounting (SAC-SMA) Model (Burnash et al. 1973) coupled to the snow accumulation and ablation Anderson Snow Model (Anderson 1973). Details on the application of these models can be found in Miller et al. (2003). The basins that were selected for this study are: the Smith River at Jedediah Smith Redwoods State Park, Sacramento River at Delta, Feather River at Oroville Dam, American River at North Fork Dam, Merced River at Pohono Bridge, and Kings River at Pine Flat Dam (Table 2.1).

The resulting streamflow climatologies based on these estimates are shown in Figure 2.3. In both GCM projections, there is a generalized tendency towards earlier streamflow peaks. This tendency is consistent with the results found in previous studies (Revelle and Waggoner 1983; Gleick 1987; Lettenmaier and Gan 1990; Jeton et al. 1996; Miller et al. 1999; Wilby and Dettinger 2000; Knowles and Cayan 2002). The effects associated with HadCM are very pronounced, with large shifts in both the amount and timing, while those associated with PCM show mainly a shift in timing and reduced magnitude. The HadCM-based streamflow shifts between 30 and 60 days earlier, and the PCM is less than or about 30 days earlier near 2100. An analysis of the changes in the snow to rain ratios, snow water equivalent, and snowmelt related to earlier streamflow timing is presented in Miller et al. (2003). Although generally the effects of the changes in streamflow volume are easier to interpret than the changes in the timing of streamflow peak, in a region with heavy reliance on reservoir storage and management of water, timing alone can have significant impacts in the operation and planning of water resources in the SJR. The following section discusses how these impacts on natural streamflow in the Central Valley translate into changes for the Valley's water management systems.

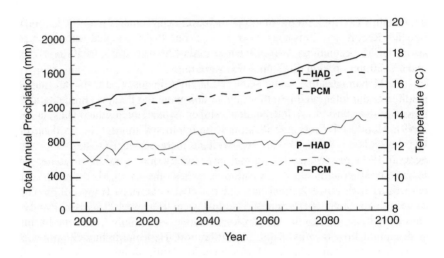

Source: Miller et al. 2003

Figure 2.2   *California mean annual temperature and precipitation projections for the 21st century for the HadCM and PCM GCMs*

Table 2.1   *Geomorphological characteristics and gage location for six key California basins*

	Smith	Sacramento	Feather	American	Merced	Kings
Area (sq. km)	1 706	1 181	9 989	950	891	4 292
Gage Latitude	41° 47'30"	40° 45'23"	39° 32'0"	38° 56'10"	37° 49'55"	36° 49'55
Gage Longitude	124° 04'30"	122° 24'58"	121° 31'00"	121° 01'22"	119° 19'25"	119° 19'25"
Percent Upper	0	27	58	37	89	72
Upper-centroid Elevation (m)	–	1 798	1 768	1 896	2 591	2 743
Lower Centroid Elevation (m)	722	1 036	1 280	960	1 676	1 067

Source: Miller et al. (2003)

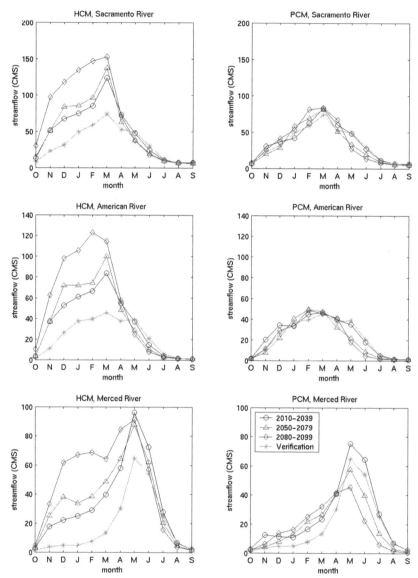

Source: Miller et al. 2003

Figure 2.3   *Monthly climatologies for historical and GCM projections of streamflow at three California Basins*

Hydrologic Impacts Based on Regional Climate Change Assumptions

In place of the GCM runs, P ratios and T differences in the six headwater basins (Table 2.1) were prescribed uniformly for all months (Table 2.2) and applied over the California climatology. These altered P and T estimates were used in the coupled SAC-SMA and Anderson Snow models. The resulting streamflow climatologies are shown in Figure 2.4. As can be seen, T shifts seem to be mainly responsible for the changes in the streamflow peak timing, while P affects the volume.

Table 2.2   The six specified incremental temperature shifts and precipitation ratios

Headwater basin	Temperature shift (°C)	Precipitation ratio
Smith	1.5	1.00
Sacramento	1.5	1.09
Feather	3.0	1.00
American	3.0	1.18
Merced	5.0	1.00
Kings	5.0	1.30

Source: Miller et al. (2003)

WATER ALLOCATION SIMULATIONS

Regardless of whether the streamflow estimates were derived from global or regional climate change assumptions, streamflow ratios were subsequently computed for each of the California basins shown in Table 2.1 by dividing the streamflow monthly climatologies from the climate change scenarios by the historical streamflow climatology of each basin. Streamflow ratios for the known basins were applied as perturbations over the streamflow time-series from similar basins draining into the reservoirs of the Sacramento River Basin and the SJRB (Brekke et al. 2004). The altered reservoir inflows were included in the water allocation model for this assessment (CALSIM II). These inflow estimates correspond to water years from 1922 to 1994, which currently serves as the critical hydrologic planning sequence for state and federal agencies in California (Brekke et al. 2004).

The water allocation model, 'CALSIM II Benchmark Study G-Model' simulates the joint monthly operations of the CVP and SWP systems.[1] CALSIM II was developed by the California Department of Water Resources and the US Bureau of Reclamation's Mid-Pacific Region Office. It includes

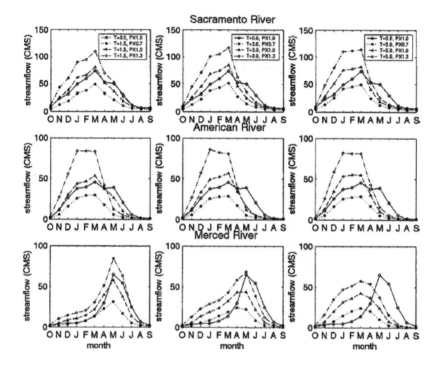

Source: Miller et al. (2003)

Figure 2.4   Monthly climatologies for historical and GCM projections of streamflow at three California basins for the incremental precipitation ratios and temperature differences shown in Table 2.2 (from Miller et al. 2003)

physical facilities for SWP, CVP and local water systems of the greater Central Valley along with codified rules governing the operations of these systems (Brekke et al. 2004). The water is allocated on a monthly basis using a mixed-integer linear programming optimizer constrained by physical constraints (such as capacity of canals and pumps) and by operation rules representing diverse water allocation contracts and water quality regulations. The objective function of the optimization is determined by a linear combination of the priorities given to meet agricultural, environmental and urban demands. State and Federal allocation priorities have significantly evolved since 1992 (Brickson 1998). CALSIM II is available in several land use/development and regulatory regime versions. The version accepted as Baseline in this study features 2001 land use assumptions and CVP/SWP operations according to the 1995 Delta Water Quality Control Plan codified in California State Water Rights Decision 1641.

Operations/Allocation Impacts: Global Climate Change Assumptions

Some of the results from the CALSIM II simulations labeled as 2025 and 2065 in relation to the HadCM and PCM climate projections are shown in Figures 2.5–2.7 (Brekke et al. 2004). The data of these figures were classified in effects during dry, normal and wet years classified according to terciles of the combined annual inflow of the Sacramento River Basin and the SJRB under each climate scenario. As can be seen, HadCM-based results show significant increases in the stored water volume in New Melones Reservoir, the main storage unit of the SJR, for scenarios centered in 2025 and 2065. In the case of the PCM, a significant reduction in the storage is expected in the 2065 climatology, but not significant in the 2025 climatology (Figure 2.5). The monthly distribution of water released by the New Melones Reservoir changed for the HadCM scenarios relative to Baseline. More water is released earlier in the year due to a shift in inflow timing and magnitude. Release increases during the period of higher demand (warm-season months) do not increase as much as the expected winter and early spring releases. The New Melones Reservoir releases for the PCM scenarios showed marginal decreases for all climate change scenarios (Figure 2.6). The water from New Melones is the most important way to meet salinity constraints at Vernalis. Presumably, this strong constraint would force CALSIM to maintain the release of water at New Melones even under decreased supply, sacrificing reservoir storage and water delivered to agriculture in the West Side SJRB (Figure 2.7).

Operations/Allocation Impacts: Regional Climate Change Assumptions

Impacts in this section include water-related operations variables of the SJR region and the greater Central Valley. The streamflow climatologies from the previous section were transformed into key water management variables in the Central Valley using CALSIM II. The resulting sensitivities are plotted in Figure 2.8.

Increases in annual outflow to the Delta due to increased P would be somewhat offset by increases in T (Figure 2.8a). This is reflected in the slope of the contours of Figure 2.8a: near-vertical lines would suggest no influence of the variable to P increases, while near-horizontal contours would suggest that the variable is insensitive to T increases. Outflow to the Delta depicts a linear influence of both P and T, with an increasing gradient of approximately +285 hm^3 per 1 percent increase in P and 1°C of T.

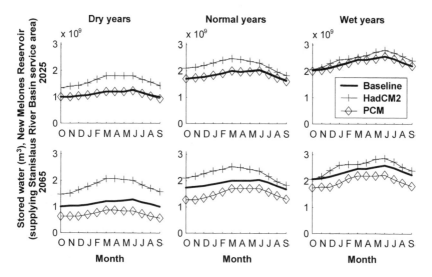

Source: Brekke et al. 2004

Figure 2.5   *Simulated monthly mean storage in New Melones Reservoir for relative Dry, Normal and Wet years for 2025 and 2065*

The percentage allocation of annual water deliveries relative to demand for one of SWP's prime contractors, the Municipal Water District (MWD) of Southern California, is also affected by both increases in T and P. However, this gradient is much smaller than the expected percent reduction in average annual allocation to the CVP agricultural contractors of the SJR West Side (Figures 2.8b and 2.8c). In terms of the April to August 'irrigation season' the CVP exports to agriculture and SWP exports to MWD through Tracy and Banks show relatively similar sensitivity to P and T increases (Figures 2.8n and 2.8o). The relative priority given to urban and agriculture water uses in this particular version of CALSIM II would have effects on how much water is allocated to the West Side and MWD users. The allocation of water to the West Side SJR for agriculture also affects return flows to the mainstem SJR, which has potential impact on the water quality of the SJR at Vernalis. This impact has to be offset by releases from reservoirs from the East Side SJR, in particular from the New Melones Reservoir. Annual agricultural deliveries from New Melones (usually to the SJR East Side) are insensitive to T changes for high P increases (P increase > 5 percent), and moderately affected by T changes for low P increases (P increase ≤ 5 percent). Annual releases to New Melones are linearly related to both P and T increases (Figure 2.8e), while the summer releases show a similar relative sensitivity as the annual agricultural deliveries from the same reservoir (not shown).

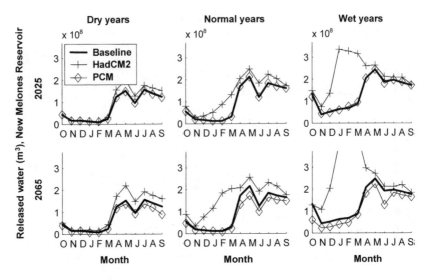

Source: Brekke et al. 2004

Figure 2.6   *Simulated monthly mean release from the New Melones*
              Reservoir to the San Joaquin River region for Dry, Normal and
              Wet years for 2025 and 2065

As was discussed in previous sections, the water quality objective at
Vernalis is a high priority constraint in the version of CALSIM II used. This
regulation is maintained to the expense of increases in the releases of
environmental flows from New Melones (Figure 2.8e). As can be seen from
the analysis, April to August EC at Vernalis is maintained in long-term
average below regulation for all simulations, even for the most unfavorable
scenario (P increase = 0 and T increase = +5°C). It should be noted however,
that mean values can be misleading in the sense that there are indeed
violations of the regulation during droughts. In fact, there is a tendency
towards more frequent violations of the regulation as temperature increases
(Figure 2.8h), on the order of 5 percent increase of the months reporting
violations of the regulations per 1°C of increase in T. In general, April to
September mean EC at Vernalis increases with increased temperature
(gradient = 25 µS cm^{-1} per 1°C) and practically insensitive to P increases.
Conversely, flow at Vernalis is strongly sensitive to P increases (gradient = 23
hm^3 per 1 percent of P increase) and not on T increases. Another measure of
the health of the San Francisco Delta and estuary is the location of the X2
point. The mean April to August location of the X2 point is strongly sensitive
to P increases (Figure 2.8p) and less related to temperature increases.

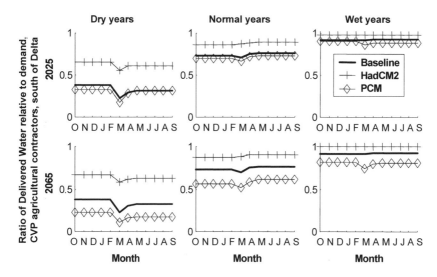

Source: Brekke et al. 2004

Figure 2.7   *Simulated monthly mean delivery levels (for example ratio of delivery to demand) for all CVP agricultural users south of the Delta, representing the West Side service areas of the San Joaquin River region, for Dry, Normal and Wet years from 2025 and 2065*

Annual allocation to the CVP agricultural contractors in the Sacramento Valley and SWP agricultural contractors in the Tulare Basin are affected considerably by both P and T increases (Figures 2.8i and 2.8j). Large reservoirs in the north of Delta basin such as Lake Oroville, Shasta, Trinity and Whiskeytown, have a sufficient storage buffer to be relatively unaffected by T increases, especially at high P levels (Figure 2.8k and 2.8m). Annual releases at Folsom and Natomas reservoirs, however, seem to be strongly sensitive to large temperature increases (increases of T greater than 2°C).

WATER QUALITY IMPACTS IN THE SJR AND DELTA REGIONS

The water quality regulations at Vernalis have to be embedded into CALSIM's logic. However, CALSIM II is not designed as a transport model. For this reason, the salinity at Vernalis is approximated through a modified Kratzer equation that relates flow and EC. It should be noted that the relation-

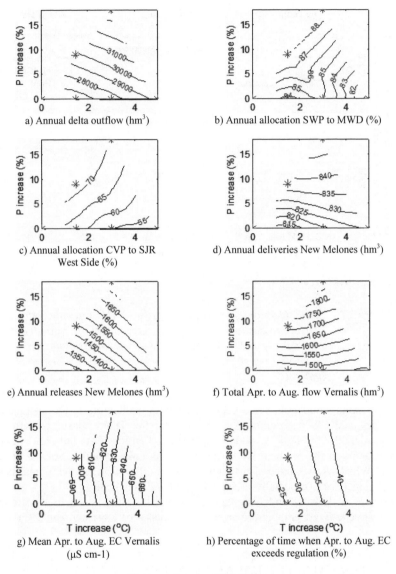

a) Annual delta outflow (hm³)

b) Annual allocation SWP to MWD (%)

c) Annual allocation CVP to SJR
West Side (%)

d) Annual deliveries New Melones (hm³)

e) Annual releases New Melones (hm³)

f) Total Apr. to Aug. flow Vernalis (hm³)

g) Mean Apr. to Aug. EC Vernalis
(µS cm-1)

h) Percentage of time when Apr. to Aug. EC
exceeds regulation (%)

(figure continued on next page)

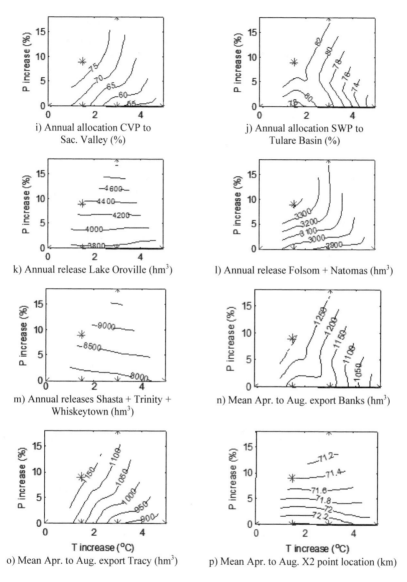

i) Annual allocation CVP to
Sac. Valley (%)

j) Annual allocation SWP to
Tulare Basin (%)

k) Annual release Lake Oroville (hm³)

l) Annual release Folsom + Natomas (hm³)

m) Annual releases Shasta + Trinity +
Whiskeytown (hm³)

n) Mean Apr. to Aug. export Banks (hm³)

o) Mean Apr. to Aug. export Tracy (hm³)

p) Mean Apr. to Aug. X2 point location (km)

Note: Asterisk symbols show the location of the points used for drawing the contours.

Figure 2.8   Sensitivity of key water-related variables to increases in annual
precipitation and temperature potentially associated with
anthropogenic climate change, assuming no land use change

ship between flow and EC at Vernalis is not evident in Figure 2.8f and 2.8g since the values of these plots are not matching pairs of monthly flow and EC values, but instead represent the average conditions of 73-years of the April to August totals (or means). This approach does not link climate-induced changes of agricultural practices, return flows and return water quality in the SJR Valley to changes in quality at Vernalis. A comparison of the measured and modeled flow and electrical conductivity at Vernalis is given by Quinn et al. (2004).

In the case of X2, CALSIM II relies on an approximation of results from an ANN model that emulates a fine-scale physical model (DSM2) of Delta hydrodynamics and water quality. The ANN has been developed by the DWR to mimic how their fine-scale hydrodynamic model (DSM2) responds to dominant inputs (such as Delta inflow and export levels, and Delta outflow) in terms of EC estimates at key Delta locations. The ANN was developed as a necessary companion pre-processor to CALSIM II because the CALSIM II model simulation requires a much simpler computation of Delta water quality than what DSM2 can afford. The CALSIM II implementation of ANN results is a lookup table that relates CALSIM-solved Delta inflow, outflow, and exports to ANN-estimated EC values given common inflow, outflow and export inputs.

SJR Water Quality

At the onset of the US EPA STAR project it was recognized that CALSIM's flow-EC relationship at Vernalis is inadequate at low San Joaquin River flows such as occur during dry and critically dry years (Quinn et al. 2004). A methodology to improve this relationship through the use of a calibrated flow and water quality hydrodynamic model (DSM2) was developed. With the linkage between CALSIM II and DSM2 in place, initial estimates of any additional New Melones water quality releases made with the CALSIM II regression were error-checked against DSM2. An iterative procedure was developed to converge on an improved release estimate.

An additional advantage of creating a linkage between the CALSIM II water allocation model and the DSM2 hydrodynamic flow and water quality model is that it allows the potential effect of any change in land use in the San Joaquin River Basin watershed to be simulated. For example, an initial working hypothesis in this project was that a drier climate signal would reduce snowpack and hence the amount of stored water behind the State and Federal reservoirs. This in turn (as simulated by CALSIM II) would lead to reduced water exports south of the Sacramento-San Joaquin Delta and a reduced water delivery to West Side agricultural water districts. Reduced water deliveries would encourage groundwater pumping and water reuse which would lead to return flows with a higher concentration of salts conveyed to the San Joaquin River.

To more comprehensively simulate the impact of reduced water deliveries on land use in the San Joaquin Basin, a model called APSIDE (Agricultural Production Salinity Irrigation Drainage Economics) was developed. The model estimates agricultural yield and productivity response to reductions in water supply, irrigation water quality, root zone and groundwater salinity, and predicts future agricultural drainage flows and water quality. Subsurface drainage flows from agricultural lands on the West Side of the San Joaquin Basin, because of the marine origin of the soils, have a significant impact on SJR water quality. In the absence of a means to export these contaminants, levels of salt and boron build up in the crop root zone and shallow groundwater aquifer reducing crop yield, leading to reduced agricultural income and eventual retirement of the land when production costs exceed farm income. The crop production function contained within the APSIDE objective function attempts to maximize land productivity by minimizing costs of agricultural production as well as salinity impacts of crop yield.

Drainage return flows generated by the APSIDE model must be routed to the SJR in order to determine the impact of these activities on river water quality. DSM2 can be used to make these computations. The model calculates the load contributed from each source based on its flow and concentration using a mass balance accounting method. The flow and water quality computation performed by DSM2, as previously recognized, is superior to the flow-salinity statistical relationship used in CALSIM II to estimate salinity concentration at Vernalis and to simulate reservoir release operations at New Melones Reservoir. Hence in instances where the 30-day running average salinity objective is exceeded, CALSIM II Stanislaus River dilution flows need to be updated as does the storage volume in New Melones Reservoir.

At the conclusion of the project, a five water district version of APSIDE had been completed. Since the water districts included in this initial version of the model were those most impacted by shallow water tables and salinity, they were considered the most likely to respond to changes in water deliveries and the return flows from these districts the most likely to show changes in water quality.

In the PCM climate scenario, where a reduction in mean monthly precipitation produces a decline in reservoir storage in the CALSIM II model, the results suggest an average annual reduction in water deliveries of about 50 percent to agriculture on the West Side of the San Joaquin Basin (Brekke et al. 2004) under the worst case scenario. Surprisingly, preliminary APSIDE results showed that reductions in water deliveries did not appear to affect the loading of salts predicted by the model to return to the San Joaquin River. Although the concentration of return flows was seen to rise over time, this appeared to be complemented by a decrease in drainage volume along with increase in the releases from New Melones, leading to very little impact on salt loads. The APSIDE model estimates production costs associated with more than twenty combinations of irrigation and drainage practices ranging from low technology, low cost options such as half-mile furrow irrigation without drainage recycling to high cost subsurface drip irrigation. The model modifies irrigation and drainage practices in circumstances where there is economic advantage. Drainage volumes predicted by the APSIDE model are dynamic, since they are the result of potential changes in agricultural land use, irrigation and drainage technology adoption and land retirement decisions over time. Initial runs of the model suggest that the agricultural system on the West Side of the Basin has a certain resilience and is able to adapt to the change in water availability, at least in the intermediate term, without significantly affecting water quality in the San Joaquin River. Future analysis using APSIDE in several irrigation districts would help verify these preliminary findings.

For this reason the linkage between the APSIDE and DSM2 models, considered essential at the start of the project, took lower priority and was not completed at the end of the project. Future work with the APSIDE model would expand the model coverage to the entire salinity-affected West side on the San Joaquin Basin and would rely on a fully coupled surface and groundwater simulation model to simulate the interchange of water between shallow and deep groundwater in the regional aquifer as well as lateral flows between adjacent water districts.

Delta Water Quality

Another water quality issue that CALSIM II must deal with is salinity intrusion into the Delta from San Francisco Bay. After years of controversy over this issue and modeling disputes between agencies, the US EPA developed a methodology for assessing the negative impacts of salt intrusion using the location of the X2 point. Since the location of X2 must be factored into the operations that guarantee a minimum Delta outflow from the Sacramento and San Joaquin Rivers exported by the projects and hence the amount of export pumping from the Delta – CALSIM II must contain logic to perform this simulation. To simulate X2, CALSIM II looks up results from an artificial neural network rendition of the DSM2 model, described at the beginning of this section, which simulates Delta hydrodynamics and water quality.

IMPACTS OF CLIMATE CHANGE ON SALMON PRODUCTION

The GCM runs from the above section on water quality under global climate change assumptions were used to assess the impact of climate change on salmon population in the SJR tributaries using an ANN model that relates streamflow and other variables to fish counts in the main SJR tributaries. An ANN is a numerical estimation scheme. Essentially, an ANN is presented with input data (for example streamflow) deemed valuable for the process being modeled (salmon populations). The ANN can 'learn' the unknown pattern between the input and output values if it is first presented with a set of training data. In this research, the neural network model was presented with three to four (depending on the basin) types of input data:

1. mean seasonal streamflow (in cfs) during salmon migration months (January–May);
2. mean seasonal streamflow during salmon spawning months (September–December);
3. annual average of the Pacific Decadal Oscillation (PDO) (Mantua et al. 1997); and
4. binary operational state (OS) variable for presence of major dam (Tuolumne River only).

The predict and for the ANN training was the annual count of spawning salmon observed in three SJRB tributaries: the Stanislaus, Tuolumne and Merced Rivers (DFG 2003). These population records covered calendar years 1952 to 2002 for the Stanislaus and Tuolumne Rivers and 1957–2002 for the Merced River.

Chinook salmon is an anadromous species. Salmons spend one to four years of their lives in the Pacific Ocean before returning to their natal stream to spawn and die (Healey 1991; Mesick 2001). Due to the variable oceanic durations, it was necessary to use five years of each predictor value to account for the cumulative variety of environments experienced by each spawning cohort. This totaled three to four variables per year for five years (for example 15–20 predictors for each year's annual count of salmon). For example, the number of Tuolumne River salmon in 1952 were calibrated using seasonal streamflow, PDO and OS data from 1948–1952.

The CALSIM II streamflow data corresponding to the GCM projections were input to the ANN for each basin, and used to project the future Chinook salmon populations. Since future values of the PDO are unknown, the future PDO values were chronological replicates of the PDO values during the training period. It was also assumed that major dams would remain on each river, so the OS variable remained equal to one during all future simulations.

Annual salmon populations were predicted for 68-year intervals, centered upon the years 2025 and 2065. Generally speaking, the phase of the 2025 and 2065 salmon time-series were found to be similar, even though their absolute numbers differed. The salmon time-series for 2025, along with the reference salmon population during the baseline years, is displayed in Figure 2.9. The annual average number of spawning salmon – in each basin and under each climate scenario – is listed in Table 2.3.

Table 2.3   Mean number of fall run Chinook salmon per basin, for historical data and different climate change projections

	Baseline	2025		2065	
	Observations	HadCM	PCM	HadCM	PCM
Stanislaus	4 111	11 721	4 293	14 415	2 906
Tuolumme	9 891	27 301	9 681	33 576	7 625
Merced	3 044	7 529	2 468	8 525	2 026
San Joaquin (sum)	17 046	46 551	16 442	56 516	12 197

Across the baseline and future periods, the fraction of San Joaquin salmon contributed by each basin remained relatively fixed: approximately 24 percent, 58 percent and 18 percent for the Stanislaus, Tuolumne and Merced basins, respectively. However, the climate models play a significant role in the projections. The HadCM estimates many more salmon (particularly in the Tuolumne and Stanislaus Rivers) than the PCM, especially with the progression of time toward 2065. This is because 2065 exhibits more of a divergence between the relatively wet forecasts of the HadCM and the drier predictions of the PCM. Since salmon respond favorably to high streamflows, the HadCM climate projections result in much higher Chinook runs. In contrast, PCM runs for 2025 do not differ appreciably from the baseline, and are noticeably lower than baseline for the 2065 horizon.

The wet and dry characterizations by the HadCM and PCM models, respectively, provide an envelope for the possible range of future salmon populations in the SJRB. To quantify the likelihood for annual salmon runs of various sizes, we have generated a quantile plot of the salmon populations based on the 2025 and 2065 ANN simulations, using future streamflow generated by the HadCM and PCM models. For comparison, the baseline period has also been included. These frequency distributions are presented in Figure 2.10.

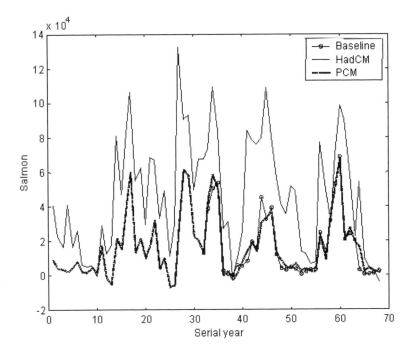

*Figure 2.9   Modeled number of spawning salmon in SJRB (all basins) for
climate model runs centered on 2025*

As shown in Figure 2.10, the PCM 2025 very closely matches the
baseline data, with PCM 2065 being lower than the baseline. The HadCM
population data are higher in all cases, with 2065 exceeding 2025. If the
HadCM simulations are correct, they indicate that Chinook salmon will
benefit from future climate change in the SJRB. If the PCM simulations are
closer to how the future climate will behave, then we expect Chinook salmon
runs to decline slightly in the SJRB. These population statistics involve
several important assumptions:

1. the operational state of each river basin does not change appreciably;
2. the range of streamflows experienced during the baseline period are
    representative of future streamflow values;
3. the future PDO is suitably represented by the PDO of the baseline period;
    and
4. oceanic factors affecting salmon do not change markedly in the future,
    such as over-fishing, aquatic disease epidemics or proliferation of
    competitive/predatory species.

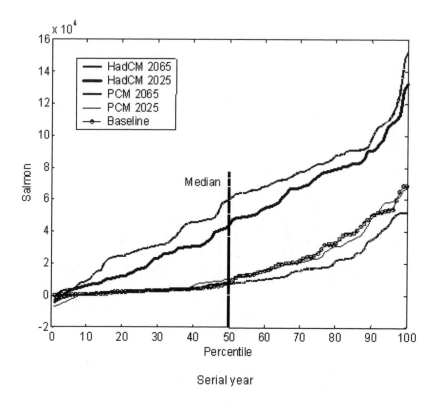

*Figure 2.10   Empirical distribution functions for the annual number of
spawning salmon in the SJRB*

CONCLUSIONS

Potential changes in the volume and distribution of streamflow associated
with climate change would have significant impacts in water supply, water
quality, and salmon runs in the SJRB. The range of the impacts from the
GCM projections incorporates the possibility that California could expect a
moderate reduction of the water available for agriculture in the SJR under the
PCM scenarios and an equal possibility of a considerable increase in winter
precipitation under the HadCM scenarios. Such a broad range of possibilities
would provide stakeholders with first-hand measure of the range of
uncertainty inherent in climate change projections and of the potential impacts
in California. Interpreting these results should remain somewhat qualitative
and focused on trends due to the overall uncertainty in model projections.

Regarding the climate projection and hydrologic impacts assessment, the approach used by Miller et al. (2003) is one of the current impact assessment approaches used by the IPCC. In future studies, dynamic down-scaling should be incorporated into these studies to determine the scale required for capturing orographically produced precipitation in California. Statistical down-scaling of GCM data will increase the resolution but will not capture orographic precipitation properly. It is also necessary to determine how many down-scaled runs are required for quantifying regional climate model variability in the region, suggesting that there should be an ensemble of down-scaled simulations, each with slightly different initial conditions, for each GCM simulation. Given the above limitations, this section of the study still provides an important and reasonable set of upper and lower bounds of hydrologic response to climate change in California. Climate models will never predict the future, but can provide projections with an uncertainty that can be bracketed. The two GCM scenarios used in this chapter represent the end members of the most-likely climate change response for California (Miller et al. 2003). Moreover, without any other information, these scenarios as well as any scenario in between have to be considered equally probable (Brekke et al. 2004). This leaves unresolved uncertainty about California's regional response to climate change, which can be simulated but ultimately reflects assumptions on global CO_2 and greenhouse gas conditions and GCM model structure to simulate global response under these conditions. Another unresolved uncertainty of regional climate change is that the changes in temperature and precipitation could vary according to the seasons (or months). It is also possible that the maximum temperatures could change differently than the minimum temperatures. These details could potentially bring large changes in the water allocation in a heavily regulated system such as the Central Valley. These considerations should be included in future studies by running more GCM scenarios.

Given a set of water allocation priorities embedded into the version of CALSIM II discussed herein, results suggest that the water quality constraints at Vernalis play an important role in the allocation of water in the SJRB. The SWRCB water quality objectives for Vernalis are met in many cases at the expense of increases in the releases from New Melones and reductions of water delivered to the CVP and SWP. In particular, CVP reductions to West Side SJR are strongly sensitive to P and T changes. The simulations also show that there is a strong sensitivity of the frequency of months in which Vernalis EC exceeds regulation with increased T (on the order of approximately 5 percent increase in the number of months reporting violation of the EC regulations per 1°C of increase in T).

Annual agricultural deliveries to the East Side are strongly sensitive to P increases. The large reservoirs in the Sacramento basin (Oroville, Shasta, Whiskeytown and Trinity) are generally insensitive to T increases, with the exception of Folsom and Natomas reservoirs that showed sensitivity to both P and T. Water exports from Tracy and Banks seem to be equally affected by P and T changes. This is particularly true for the exports from the SWP to the Tulare Basin and Metropolitan Water District. Outflow to the Delta depicts a linear influence of both P and T, with an increasing gradient of approximately +285 hm^3 per 1 percent increase in P and 1°C of T. The location of the X2 point showed little sensitivity to T, and is mainly responsive to P increases.

Salmon populations varied according to the simulation year and the GCM projection used. The HadCM presents a wetter future climate, with significantly more Chinook salmon. The PCM foresees a drier climate than today, with a slight reduction (6 percent) in the median number of spawning salmon. Effectively, these two general circulation models produce an envelope for the range of future Chinook salmon populations in the SJRB, provided that external influences such as oceanic, biological and anthropogenic factors remain constant.

Future adaptation and mitigation climate change studies in California will require strategies that affect the combination of variables in a complex system with regulated and unregulated components. Modifications of the unregulated component are represented by significant alterations in the hydro-climatological regime in the headwaters, principally affecting wintertime water supply, but also in the changes in warm-season evapotranspiration in the Valley (Hidalgo et al. 2005). Changes in the managed system are related to different operational strategies in response to altered patterns of water supply and demand. Such strategies involve modifying the land use (for example changing crops or removing land out of production), improving agricultural practices (such as through technology), modifying the operation of the reservoir or canals systems, and proposing adjustments to environmental regulations and water allocation priorities meant to preserve Central Valley and Delta environmental resources.

Benefit and cost analysis of many sets of strategies should be estimated. In order to do this, it is necessary to model the agro-economics of the SJRB along with the climate and water quality projections in order to estimate future land use changes under altered climate conditions. At the very least a reliable irrigation–drainage–groundwater model compatible with CALSIM is necessary. Efforts are underway to produce reliable models for agricultural production and agro-economics in the SJR.

In some cases, it may be impractical to coordinate individual (and sometimes incompatible) models for water allocation, water quality, water temperature, agricultural production, irrigation–drainage, groundwater and economics. One alternative is to run the models in serial, using the output to one as input to the other(s). This has the advantage of providing detail of the impacts in some regions, while allowing other models to be more general in scope. The disadvantage is that feedbacks between the models may be difficult to identify or to incorporate in the modeling process. Another alternative is to use the information from the individual models to incorporate a simplified version of some of these aspects into CALSIM II logic or another model. This would allow the analysis of a greater number of mitigation/adaptation strategies and GCM runs. It has the disadvantage of requiring more data from regions out of the area of interest. Efforts to collect, archive and make long-term water quality data available should be supported in order to produce reliable models and improve our assessments of climate change impacts, and to extend our planning horizons.

ACKNOWLEDGMENT

This research has been funded by the United States Environmental Protection Agency through the Science to Achieve Results program (Grant R827448).

NOTE

1. Details on CALSIM II and other modeling tactics can be found through the California Department of Water Resources, Bay-Delta Office on-line at http://modeling.water.ca.gov.

REFERENCES

Anderson, E.A. (1973), *National Weather Service River Forecast System: Snow Accumulation and Ablation Model*, NOAA Technical Memorandum NWS HYDRO-17.

Brekke, L.D, N.L. Miller, K.E. Bashford, N.W.T. Quinn, and J.A. Dracup (2004), 'Climate change impacts uncertainty for water resources in the San Joaquin River Basin', *Journal of American Water Resources Association,* **40**: 149–64.

Brickson, B. (1998), *Layperson's Guide to the Central Valley Project*, Sacramento, CA: Water Education Foundation.

Brown, R. and W. Kimmerer (2001), *Delta Smelt and CALFED's Environmental Water Account*, Summary of a Workshop held 7 September 2001, Putah Creek Lodge, Prepared for the CALSIM Science program, University of California, Davis.

Burnash, R.J.C., R.L. Ferral and R.A. McQuire (1973), 'A generalized streamflow simulation system', in *Conceptual Modeling for Digital Computers*, Silver Spring, MD: US National Weather Service.

Daly, C., T.G.F. Kittel, A. McNab, J.A. Royle, W.P. Gibson, T. Parzybok, N. Rosenbloom, G.H. Taylor and H. Fisher (1999), 'Development of a 102–year high resolution climate data set for the conterminous United States', in *Proceedings, 10th Symposium on Global Change Studies, 79th Annual Meeting of the American Meteorological Society*, Dallas, Texas, pp. 480–83.

Department of Fish and Game (DFG) (2003), *'Grandtab' Data Set of Fall–Run Chinook Salmon Spawner Populations*, California Department of Fish and Game, Provided by B. Kano, 20 February 2003.

Department of Water Resources (DWR) (1998), *California Water Plan, Volume I*, Sacramento, CA: California Department of Water Resources (Bulletin 160-98).

Dracup, J.A. and S.D. Pelmulder (1993), 'Estimation of monthly average streamflow for the Sacramento–San Joaquin River system under normal, drought, and climate change conditions', in *Integrated Modeling of Drought and Global Warming: Impacts on Selected California Resources*, Report by the National Institute for Global Environmental Change, University of California, Davis.

Gleick, P.H. (1987), 'The development and testing of a water balance model for climate impact assessment: modeling the Sacramento Basin', *Water Resources Research*, **23**: 1049–61.

Healey, M.C. (1991), 'Life history of Chinook salmon (*Oncorynchus tschawytscha*)' in C. Groot and L. Margolis (eds), *Pacific Salmon Life Histories*, Vancouver, BC: UBC Press, pp. 311–93.

Hidalgo, H.G., D.R. Cayan and M.D. Dettinger (2005), 'Sources of variability of daily evapotranspiration in California', *Journal of Hydrometerology*, **6**: 3–19.

Intergovernmental Panel on Climate Change (IPCC) (2001), *Climate Change 2001: The Scientific Basis, Intergovernmental Panel on Climate Change*, Cambridge: Cambridge University Press.

Jeton, A.E., M.D. Dettinger and J.L. Smith (1996), *Potential Effects of Climate Change on Streamflow: Eastern and Western Slopes of the Sierra Nevada, California and Nevada*, US Geological Survey, Water Resources Investigations Report, pp. 95–4260.

Knowles, N. and D.R. Cayan (2002), 'Potential effects of global warming on the Sacramento/San Joaquin Watershed and the San Francisco Estuary', *Geophysical Research Letters*, **29**: 38-1–38-4.

Kuminoff, N.V., D.A. Sumner and G. Goldman (2000), *The Measure of California Agriculture, 2000*, Published by the University of California Agricultural Issues Center, Revised, updated and expanded from H.O.

Carter and G. Goldman, *The Measure of California Agriculture: Its Impact on the State Economy,* December 1998, Davis, CA.

Lettenmaier, D.P. and T.Y. Gan (1990), 'Hydrologic sensitivities of the Sacramento–San Joaquin River Basin, California, to global warming', *Water Resources Research,* **26**: 69–86.

Mantua, N.J., S.R. Hare, Y. Zhang, J.M. Wallace and R.C. Francis (1997), 'A Pacific interdecadal climate oscillation with impacts on salmon production', *Bulletin of the American Meteorological Society,* **78**: 1069–79.

Mesick, C. (2001), *Factors That Potentially Limit the Populations of Fall-run Chinook Salmon in the SJR Tributaries,* Unpublished manuscript.

Miller, N.L., J. Kim, R.K. Hartman and J. Farrara (1999), 'Downscaled climate and streamflow study of the southwestern United States', *Journal of American Water Resources Association,* **35**: 1525–37.

Miller, N.L., K. Bashford and E. Strem (2003), 'Potential climate change impacts on California hydrology', *Journal of American Water Resources Association,* **39**: 771–84.

Quinn, N.W.T., L.D. Brekke, N.L. Miller, T. Heinzer, H. Hidalgo and J.A. Dracup (2004), 'Model integration for assessing future hydroclimate impacts on water resources, agricultural production and environmental quality in the San Joaquin Basin, California', *Environmental Modeling and Software,* **19**: 305–16.

Revelle, R.R. and P.E. Waggoner (1983), 'Effects of a Carbon Dioxide Induced Climatic Change on Water Supplies in the Western United States', in *Changing Climate,* Washington, DC: National Academy of Sciences Press.

US Bureau of Reclamation (US BR) (1991), *Evaluation of Central Valley Project Water Supply and Delivery Systems,* Technical Report by the USBR Mid-Pacific Regional Office, Sacramento, CA.

US Environmental Protection Agency (US EPA) (1997), *Climate Change and California* (Report No. 230-F-97-008e), Washington, DC: US EPA Office of Policy, Planning and Evaluation.

Wilby, R.L. and M.D. Dettinger (2000), 'Streamflow changes in the Sierra Nevada, California, simulated using statistically downscaled general circulation model output', in *Linking Climate Change to Land Surface Change,* S. McLaren and D. Kniveton (eds), the Netherlands: Kluwer Academic Publishers, pp. 99–121.

3 Modeling Interactions Among Wildland Fire, Climate and Society in the Context of Climatic Variability and Change in the Southwest US

B. Morehouse, G. Christopherson, M. Crimmins, B. Orr, J. Overpeck, T. Swetnam and S. Yool

INTRODUCTION

Between 1985 and 2004 wildland fires burned more than 75 million acres across the United States (NIFC 2004a, 2004b). Moreover, in 2002 the direct cost of fighting fires reached a high of more than $1.6 billion (NIFC 2004b). Damage and destruction of homes, infrastructure and ecosystems have likewise been skyrocketing. Indeed, concern about wildland fire has reached the highest levels of government (White House 2002; US Congress 2003) and accounts of dramatic fire events have become a staple of national, regional and local news media. The raging fires that occurred in southern California in early fall 2003, for example, captured sustained attention from reporters and viewers alike. In part, contemporary fire problems stem from almost 100 years of active and aggressive fire suppression. Rapid exurban development of areas near and within the region's forests has exacerbated these problems. The concern expressed at all levels of government, from local to federal, about both the impacts and costs of these fires is providing unprecedented opportunities to combine scientific expertise with on-the-ground knowledge held by fire fighters, forest managers and local communities to improve management of fire-adapted landscapes.

At the same time that fire risk is increasing, our knowledge about fire and its role in wildland ecosystems is also increasing – as is our understanding of the processes influencing geographical and interannual variability in fire regimes. This increase in knowledge has been paralleled by intensified efforts to translate knowledge into decision support tools that decision makers and fire managers can use to anticipate and manage fire risk more effectively. Such tools provide rich sources of information for determining how best to reduce destruction from catastrophic fires while at the same time allowing fire, where appropriate, to play its natural role in ecological processes.

Fire as a natural part of ecological processes is especially prominent in large parts of the western US where some landscapes, such as those dominated by ponderosa pine, are adapted to and require periodic episodes of burning. Much has been learned about the role of fire, although much fundamental and applied scientific work remains to be done. For example, knowledge has been gained as to what factors influence forest health (Covington and Moore 1994a; Covington et al. 2001), the biophysical influences on natural fire regimes and influences of fire on biophysical conditions (Baisan and Swetnam 1990; Covington and Moore 1994b; Yool 2000; Henry and Yool 2002; McHugh et al. 2003). Research has also illuminated the importance of societal factors (Baker 2001; Pyne 1982, 2001; Pyne et al. 1996).

Integrated research that provides new insights into the interactions between fire, biophysical and societal dynamics is one of the areas where new research holds particular promise for aiding decision making (see for example Conard et al. 2001; White 2004). The southwestern US constitutes an important venue for such research; and projects focused on the southwestern region, in turn, hold promise for improving wildland fire management across the US.

Climate patterns in the Southwest, for example, are known to be important factors in fire regime variability at seasonal to interannual and interdecadal time scales (Brown and Comrie 2002; Comrie and Broyles 2002; Crimmins and Comrie 2004; Sheppard et al. 2002; Simard et al. 1985). In addition, tree-ring research indicates a statistical correlation between interannual shifts in precipitation regimes and patterns of widespread fire occurrence at the regional scale across much of the West, including the Southwest (Grissino-Mayer and Swetnam 2000; Swetnam and Betancourt 1990; Baisan and Swetnam 1990). Much of the vegetation is fire-adapted, and indeed, as noted above, some species require fire to propagate. Research into vegetation dynamics (Moran et al. 1994; Running et al.1989; Covington et al. 2001; Covington and Moore 2004a) provides a link between climate and ecological dynamics, while advanced geographic information science (GIS) and remote sensing constitute essential technologies for improving knowledge of vegetation distribution and class, as well as for understanding interactions between climate, weather and fuel conditions (see for example Keane et al. 2001; Reed et al. 1994; Nemani et al. 1993; Morgan et al. 2001). When biophysical information is integrated with socioeconomic data and with geospatially referenced data on human values, possibilities arise to create innovative products useful for decision making in fire-prone landscapes by experts and by the general public as well.

CLIMATE IN THE SOUTHWEST US

Climate is an important factor in seasonal to longer-term patterns of wildland fire occurrence. The climate of the southwestern US, although generally semiarid, is characterized by a bimodal precipitation regime, with rains occurring during summer and winter (Sheppard et al. 2002). Spring tends to be predictably dry, as does fall except for occasional rain associated with passing events such as tropical storms. The El Niño Southern-Oscillation (ENSO) strongly influences interannual variability in the region, and research suggests that the longer-term Pacific Decadal Oscillation (PDO) intensifies ENSO conditions when the positive and negative phases of the two coincide. Winters in the region have a tendency to be wetter when El Niños occur and, more predictably, drier when La Niñas occur. Under certain conditions, the PDO may influence these tendencies. For example, when ENSO and PDO are both in negative phase, winter precipitation in the Southwest trends toward even drier conditions than usual. ENSO-neutral years feature a more scattered pattern of wet and dry winters. An outgrowth in scientific understanding of ENSO influences on regional climatic patterns has led to a marked improvement in winter precipitation forecasts (Hartmann et al. 2002). During years when Pacific Ocean temperatures and air pressure along the equator indicate formation of moderate to strong ENSO conditions, forecasts tend to be relatively accurate for the Southwest and for other areas such as the Pacific Northwest (where the influences of El Niño and La Niña, respectively, are opposite those of the Southwest). Although predicting precipitation during ENSO-neutral conditions tends not to be as well-developed, the knowledge that an ENSO-neutral winter is anticipated can be useful information. For wildland fire management in the Southwest, precipitation conditions over multiple years, especially those where a wet El Niño winter is followed by one or more years of anomalously dry winters, have significant implications for fuel loads, fuel moisture conditions and ultimately fire risk (Swetnam and Betancourt 1990; Grissino-Mayer and Swetnam 2000).

Temperature conditions are also important to understanding the climate of the Southwest, particularly during spring and fall, the major fire seasons. Interannual temperature variability on average may not be as marked as that of precipitation, but early onset of hot weather and/or delayed relief from high temperatures in the fall can have significant impacts on the region and its resources. Furthermore, scientific evidence is strong that temperatures in the region have increased markedly in the last couple of decades, suggesting that longer-term climatic change may be underway (Sheppard et al. 2002;

Overpeck et al. 1990). Long hot spells and the general trend toward an increase in temperatures are particularly important influences on moisture conditions in dead and live fuels and, more generally, the availability of water at times when it is most needed such as forage needed by livestock. For fire management, higher average temperatures can lead to less moisture availability in the atmosphere as well as in soils and fuels, thus exacerbating fire risk even in cases where precipitation patterns remain at or above historical norms.

Research on climate change impacts suggests that the Southwest is already experiencing an elevation in temperatures, and that future change can be expected (Overpeck et al. 1990). However, modeling climatic change in the Southwest remains problematic in no small part because the topography of the region is quite complex, and because even nested regional models do not yet do a good job of representing the North American Monsoon, which is responsible for the area's summer rainfall regime (Sprigg and Hinckley 2000). Output of the Canadian Climate Model, for example, has indicated that in the Southwest US the primary changes would be increased temperatures and increased precipitation. However, as noted above, an increase in precipitation, accompanied as it would be by increased evapotranspiration, would not necessarily result in moister conditions or diminished fire risk. An initiative currently underway to gather more data on climatic processes and conditions associated with the North American Monsoon will certainly contribute to better climate change modeling. Enhanced understanding of the implications of climatic change for ENSO and PDO processes will also be very useful. In the meantime, development of models such the one described in this chapter that focus more specifically on climatic variability at regional or sub-regional scales hold promise for addressing persistent societal and ecological problems, such as wildland fire, that are influenced at least in part by climatic processes.

WILDLAND FIRE CONDITIONS IN THE US AND IN THE SOUTHWEST US

Forty million acres of National Forest lands alone are currently in elevated fire hazard condition; these conditions prevail in much of the forested land in the Southwest US. High fuel load levels, fire suppression policies, human activities and climatic conditions all contribute to the hazard. The cost of fighting wildfires is increasing, as are wildfire-related fatalities. Given existing conditions, experts expect wildfires to last longer, encompass more acres and involve more regions of the US. An assessment issued by the National Interagency Fire Center's Predictive Services Group early in 2004 cited the persistence of long-term drought over much of the interior west, combined with insect infestations, as major reasons to be concerned about the potential for large destructive wildfires in mid and high elevations (NIFC 2004a). Another report suggests that, using data from 1994, 1996, 2000 and 2002, the National Interagency Coordination Center could expect worst-case conditions, on average, to produce 85 000 reported fires, and for more than 6.1 million acres to be burned (NIFC 2004b). The extent of the fire problem in wildland areas shows up in the statistics for acres burned during the recent past:

1.   2000 = 7 383 493 acres
2.   2001 = 3 570 911 acres
3.   2002 = 7 184 712 acres
4.   2003 = 3 959 223 acres.

The five year average for amount of land burned, through 2003, amounts to 5 477 830 acres, and the ten year average is 4 455 593 acres. In the Southwest, years such as 1994 (563 696 acres burned), 2000 (601 670 acres burned) and 2002 (1 117 993 acres burned) stand out. Based on data from 1994, 1996, 2000 and 2002, the worst fire years averaged 656 091 acres burned. At the same time, the cost of suppressing fires has become a matter of political concern. Indeed, the years 2000, 2001 and 2002 registered the three highest expenditure levels for the entire period, topping $1.6 billion in 2002. Injuries, loss of life and property damage add even greater costs to the effort of managing wildland fire.

Fire hazard conditions currently threatening valuable lands and resources in the US exist within a larger context of increasing linkages between urban growth and recreational land use (the urban–wildland interface), and complex institutional factors such as the requirements of the Endangered Species Act, the National Environmental Policy Act (NEPA), the Wilderness Preservation Act, and discussions about the positive and negative effects of prohibiting road construction in wildland areas. Managing fire in natural areas within this context is challenging even under optimal conditions. One of the primary challenges is that gaps persist in scientific knowledge about the interrelationships between human activities, climatic factors and fire frequency, extent and intensity. At the same time, new federal wildfire policy requires fire managers to plan for time periods of up to a century into the future.

Recognition of the positive role played by fire in promoting forest health has grown, reaching the highest levels of decision making and governance (Brown 1985; Baker and Kipfmueller 2001; Covington and Moore 1994a; Kolb et al. 1994; Pyne et al. 1996). Along with this recognition have come stronger efforts to include fire use in fire planning and management. Likewise, interest is growing in bringing the best available scientific knowledge about factors influencing fire regimes, as well as about fire behavior and fire impacts to seasonal and longer-term planning processes. As noted earlier, addressing the need for such information requires innovative approaches that combine the best scientific knowledge available from the biophysical and social sciences into decision support tools that are both useful and usable to the constituencies for which they are designed (Nicholson et al. 2002). FCS-1, developed at the University of Arizona under the interdisciplinary Wildfire Alternatives (WALTER) initiative and funded by the US EPA STAR Program under Grant Number R-82873201-0, is an example of the kinds of decision support research that are currently underway.

FCS-1: AN INTEGRATED MODEL FOR WILDLAND FIRE MANAGEMENT

FCS-1 is a first-generation integrated GIS model for use in *strategic* planning for wildland fire management (Figure 3.1). The name of the model, Fire–climate–society, Version 1, reflects the emphasis that has been placed on including a wider range of factors than is ordinarily included in fire models, such as BEHAVE and FlamMap, that are designed for shorter-term *tactical* use. The 'version 1' tag indicates that this model constitutes a basic framework for representing fire risk, but that it does not have all the components that ideally would be included. For example, a more fully elaborated model would have dynamic components allowing for streaming of real-time climate data and for linkage with a fine-scale vegetation model.

FCS-1 is somewhat unique in emphasizing accessibility, utility and usability for community members as well as for fire managers, fire scientists and decision makers. The model is explicitly designed for use in decision making at time scales of one month to a season, a year or longer. Based on user selection of a climate scenario and user weighting of model layers, the system provides fire hazard and fire risk maps at a grid scale of 1 km. From the beginning, the model has been envisioned as operating via internet interface, and this decision led to dedication of considerable resources and time to building a multifunctional and user-friendly website.[1]

FCS-1 identifies geographically explicit levels of risk of large wildland fire based on research, which indicates that fires growing to 250 acres are most likely to become large, destructive wildfires. The model currently encompasses four well-defined study areas: the sky island ecosystems of the Catalina-Rincon, Huachuca and Chiricahua Mountains in southeastern Arizona and the Madrean ecosystem of the Jemez Mountains in New Mexico (Figure 3.2).

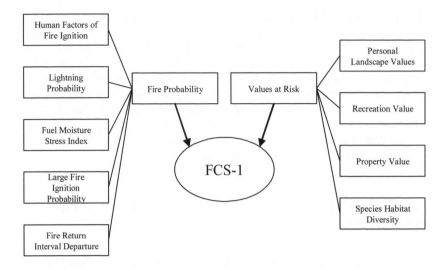

Figure 3.1   Schematic of FCS-1 model

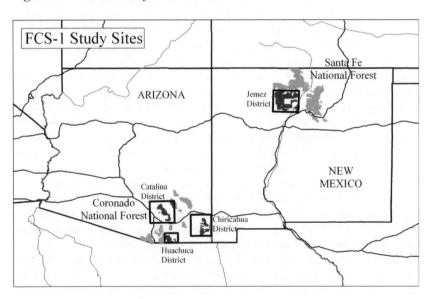

Figure 3.2   FCS-1 study sites

These four study sites were selected for several reasons. First, the principal investigators already had conducted research and had data on the areas that would provide foundations for the modeling effort. Second, each of the sites represented key types of situations faced in wildland fire management. The Catalina-Rincon Mountains border the growing city of Tucson, Arizona, and feature a large urban-wildland interface area. The Huachuca Mountains are located on the fringe of the small city of Sierra Vista, and are partially encompassed by Fort Huachuca Army Base, which is a major US military installation. These mountains also extend to the US–Mexico border and are currently being heavily impacted by illegal migrants. Further, the area is widely famous for its flora and fauna, especially the hummingbirds that are protected in the Nature Conservancy's Ramsey Canyon Preserve. It hosts large numbers of visitors all year round. The Chiricahua Mountains, located just inside the border with New Mexico and just north of the US–Mexico border, constitute the most remote and undeveloped of the study sites. Yet marketing is already underway to convert shrublands at the mountains' base to summer homes. Chiricahua National Monument, a beautiful area that encompasses important geological formations and ecological sites, is located on top of the mountain range. The mountains and surrounding areas support significant ranching activities as well as recreational use. This mountain range is the only one of the study areas that does not have a direct urban–wildland interface. As such, it provides a baseline case for modeling fire–climate–society dynamics in a significantly less impacted area. The fourth study site, the Jemez Mountains in northern New Mexico, being part of the Madrean system, comprises a somewhat different, but analogous landscape. The Los Alamos Nuclear Laboratory and the city of Los Alamos constitute major social imprints on the landscape. A few other communities, such as Espanola, are also scattered around the base of the mountain range. On the ridge of the mountains, Bandelier National Monument encompasses some of the most important Native American ruins in the area. Here again, the urban–wildland intermix is critical to fire management. Also crucial to decision making is the existence of Native American lands belonging to Santa Clara and Jemez Pueblos, and sacred sites important to these pueblos, as well as Cochiti Pueblo and others. The Laboratory is a key player in local forest and fire management activities, not only with regard to the high-security nature of the operation and the need to actively manage the forest on its lands, but also due to its energetic participation in the Interagency Wildfire Management Team, which meets every two weeks to coordinate planning and operational activities. Three of the four study areas have experienced major fires over the past 30 years, with the most recent occurring in June–July 2003 on top of Mount Lemmon in the Santa Catalina Mountains; fire forced evacuation of Los Alamos in 2000.

FCS-1 MODEL DESIGN

FCS-1, while specifically incorporating data for these four study areas, is designed with sufficient flexibility to be transferred to other forest areas in the Southwest, and to other parts of the West where similar conditions prevail. The model comprises two sub-models: the fire probability sub-model combines biophysical data layers to produce a fire hazard map, and the values at risk sub-model combines societal value data layers to produce a map of societal values vulnerable to damage or destruction by wildland fire. FCS-1 combines the layers from both sub-models to produce an integrated fire risk map. Each of the layers incorporated in the model represents the best science and data available at the time they were created. In most cases, data for the past 20 years were used. In some instances, such as lightning data and human values, it was necessary to use shorter time periods.

Fire Probability Sub-model

FCS-1 contains five data layers representing biophysical conditions contributing to fire hazard: human ignition probability, fuel moisture stress index, lightning probability, fire return interval departure, and large fire probability. These layers incorporate, as appropriate, basic characteristics such as topography, slope, aspect, roads, administrative boundaries and other features.

The *fuel moisture stress index* layer is cued by climate scenarios from which the user must make a selection. This is the only layer that allows such user input. The data making up this layer includes vegetation type and fuel moisture conditions calculated at a scale of 1 kilometer using NDVI data (ground-truthed through field research) for each selected study area, in combination with the climate scenario data. The scenario data include precipitation for the previous winter and temperature during the fire season of the scenario year selected. FCS-1 applies the indexed data to each of the pixels on the map of the selected study site.

The *fire return interval departure (FRID)* layer is based on 20 years of fire history data. The layer contains spatially explicit information about vegetation types, fire perimeters, dates of the fires, length of time since last fire, expected fire return interval based on longer-term records and vegetation type. FCS-1 calculates the FRID for each pixel using the following equation: (years since last fire – natural fire return interval) / natural fire return interval. The normalized index derived from this equation is applied to each of the one-kilometer grids to produce the FRID maps for the four study sites.

The *large fire ignition probability* component provides input regarding the likelihood that, based on vegetation type, a fire greater than 250 acres will occur within each pixel. Research that suggests that fires reaching 250 acres in size have an increased probability of growing into fires consuming more than 1,000 acres. Burn statistics and vegetation data constitute the foundational elements in this component.

The relative importance of natural versus human sources of fire ignitions reflects the likelihood of fire in specific areas based on lightning events or human activities. The *lightning ignition probability* layer identifies, by 1 kilometer grid cell, the probability that lightning will occur in that cell. The data for this layer are derived from records of lighting strikes in each of the study sites. These data were purchased from a private firm specifically for use in this model. As good lightning data are only available for the past ten years for all four study sites, only ten years of data are included in this layer.

The *human factors of fire ignitions* component represents the locations of human-set fires that have occurred in the study area. This layer is based on proxies such as distance from roads and assumes that physical factors such as road density and distance to picnic areas are reasonable representations of relationships between fire occurrence and human activities.

Values at Risk Sub-model

There are four societal values layers included in the model: personal landscape values, property values, recreation values and species habitat diversity values. The *recreation value* layer is based on calculations of Euclidean distances from roads and hiking trails to scenic vista points, campground locations, etc., on the assumption that landscape views are of considerable value to those who make the effort to go into these mountain ranges.

The *species habitat richness* layer represents a proxy for values individuals hold with regard to the existence of wildlife in the area. The data reflect the extent to which existing habitat conditions could support mammals, amphibians, reptiles and birds that might be expected to visit or reside in each pixel. This proxy was chosen due to the lack of dependable data about actual existence/distribution of different species in the study areas. Data for this layer were obtained Arizona and New Mexico GAP databases.

The *property value* layer reflects the values of owner-occupied houses within and surrounding each study site. These data were obtained from US Census files. The *personal landscape values* layer represents the results of a series of interviews in the four study areas. A total of 120 individuals were interviewed. Each interviewee was asked to mark, on a large-sized, specially designed topographic map of their study area, their responses to a series of questions. These questions included the areas they most anticipate will burn in the next five years, the area they would most hate to see burn, the areas where they have engaged in recreation over the past year, and the routes they take to get to these places. Digitized versions of these individual maps provide the aggregated base data for this layer.

User Weighting of FCS-1 Layers and Model Outputs

FCS-1, as a decision support tool for strategic planning in the realm of wildland fire, is unique in allowing users to weight the model layers themselves. They may also elect to run the model based on an expert weighting scheme that is provided. The weightings are accomplished using Analytic Hierarchy Process (AHP) (Saaty 1980, 1990; see also Schmoldt and Peterson 2000). AHP is a structured method, based on matrix algebra, that facilitates analysis of complex issues by sorting key factors into a series of pair-wise comparisons. To assign weights for one of the selected mountain ranges, the user first selects a climate scenario from the alternatives provided on the website, then selects the option allowing user assignment of weightings. The model displays the first two layers of the fire probability sub-model. The user begins by designating the relative level of importance (using a scale of 1 to 10) to the fuel moisture hazard layer relative to the FRID layer. If the user assigns, for example, a weight of 6 to this layer, it means that this user considers fuel moisture hazard to be six times more critical to fire probability than the FRID. The user would then go on to assign weights for fuel moisture hazard against vegetation type hazard, lightning hazard and human ignition hazard. Next, the user would weight FRID relative to vegetation type hazard, lightning hazard and so on. This process continues until every possible pair-wise comparison has been made for the fire probability sub-model. The user then repeats the process for all the layers in the values at risk sub-model. After this process is complete, the user assigns weights to the fire probability and values at risk sub-models in a final pair-wise comparison. Once all of the comparisons have been completed, the model calculates the results and displays them as map outputs, gridded at a scale of one square kilometer.

As is evident from the above description, the model produces multiple outputs with a wide range of potential variability. Maps for individual users may be produced, or aggregate maps representing everyone participating in the weighting exercise. The model may be used by a single expert, for example, to assess fire risk for budgeting or preseason planning for fire season staffing. At the same time, its flexibility allows use in a group setting where, for example, the intent might be to identify, discuss and perhaps resolve issues in specific geographical areas where contention exists among different interest groups.

When combined with other information provided on the WALTER website, such as animated AVHRR greenness maps, additional fire history data, fire policy information and climate information, FCS-1 constitutes a forward-looking addition to the array of decision support tools available to fire managers and – relatively unique in this case – to the public as well.

HORIZONTAL AND VERTICAL INTEGRATION: OPPORTUNITIES AND CHALLENGES

Designing decision support tools that foster horizontal integration across different disciplines, interest groups and segments of society is a challenging endeavor, but one that can have high payoffs in terms of return on investment of funds, time and energy (see Courtney 2001; Bonczek et al. 1981). The effort can also have high payoffs in terms of providing access to the best information available to members of society who wish to participate in decision processes and/or who stand to be affected by the decisions of others. Further, such tools afford opportunities for bringing together disparate sectors of society who might not ordinarily have the same access to information. However, in many cases building tools that satisfy the needs of experts and at the same time afford decision support to non-expert members of society – especially those who stand to be most strongly affected by decision making processes – is challenging under even the most ideal circumstances. As described below, the process of developing and testing FCS-1 was strongly focused on bridging the gap between experts and non-experts.

Likewise, vertical integration within modeling efforts can produce both successes and challenges. FCS-1, for example, like many models, is built with the best information that could be obtained; however few datasets and sources of information can fill all of the needs of such modeling efforts. As discussed below, the data used in the model's layers have been assessed in terms of quality assurance, and this information has been made available to the individuals who participated in workshops to evaluate the model.

Horizontal Integration

FCS-1 is explicitly designed for use both by fire management experts and by communities in the four study sites. The primary factor facilitating both public and expert use of the model is AHP: each individual is empowered to weight the model's data layers according to their own knowledge and preferences. The model is designed to allow each individual to see his/her own maps for each data layer, the composite maps for each of the sub-modules and the final combined fire risk map. Making the model accessible in this manner to both experts and non-experts has been an important goal of the project. Holding user evaluation sessions in each study area that explicitly included both expert and community participants provided an invaluable opportunity to obtain constructive feedback about how well the model performed, how well it met user needs and what adjustments – including addition of other or alternative datasets – might make it more accurate and useful. Equally important, these sessions provided an opportunity to assess success toward the goal of building a model that actually served both expert and general public needs in terms of the usefulness of the model outputs and the usability of the model itself.

Creating a model that serves both expert and public needs through horizontal integration has posed a number of challenges. One of the biggest challenges has been building the interfaces required to run the model on the web. While concepts of how the interface would look and function were developed early in the project, the need to spend the first two years developing the underlying data for the model meant that much of the intensive work on combining the various elements and ironing out the interface challenges could not be tackled until relatively late in the project funding period. Another challenge, discussed below under 'Vertical Integration', was determining how to weight the data layers of the model in a way that would produce good fire risk maps. The decision to shift this determination to the users themselves, via AHP, constituted a significant break-through in the model design and construction process. The user evaluation sessions, held at each of the study sites, affirmed that this approach of allowing users to make their own choices about the importance of the layers relative to each other was the best design strategy.

An important element in horizontal integration is the ability to run the model any number of times, and to see how different weightings and climate scenarios affect the patterns appearing on the FCS-1 fire risk maps. Also important is the possibility of integrating use of the model in group contexts. Having alternative risk maps available can facilitate discussion of points of agreement and disagreement about the level and spatial distribution of fire probability, the potential implications for values at risk and overall fire risk in the area. The group setting can also provide a venue for discussion of individual opinions about how the data layers should be weighted relative to each other and how different weightings influence model outputs. As the FCS-1 testing and evaluation sessions with stakeholders have revealed, group sessions provide an opportunity for experts and residents to analyze jointly the final map as well as maps produced by different individuals. This can encourage discussion not only about the type, quantity, quality and sources of the underlying data, but also about individuals' knowledge, fundamental values and concerns. Since the purpose of the system is to facilitate strategic planning for fire management – that is, planning a season to a year or more in advance – group sessions like these can provide opportunities to work toward forest/fire management plans that rest on a broader base of community support, and that reflect more closely the intent of policies such as NEPA, the Air and Water Quality Acts, the Endangered Species Act and so on.

Opportunities and challenges also arose in the narrower world of designing a model that would be accepted and used by forest/fire management experts. WALTER team members recognized from the beginning that expert knowledge of the local areas was a crucial factor in strategic planning. A pivotal strategy in the development of the model was direct interaction with key individuals over the course of the model construction and testing/evaluation processes to obtain the best available scientific data for each of the layers, to build the kinds of professional relationships that would ensure that the model was as useful and usable as possible, and ultimately to facilitate acceptance and use of the model. The devastating fire seasons of 2000, 2001 and 2002, together with heightened interest at federal and state levels in reducing wildland fire risk, provided an excellent opportunity for productive interaction with managers and scientific experts in the development of FCS-1. Through cooperation with these individuals important datasets were obtained, as was early feedback on the model's design and components. The evaluation/testing sessions proved especially useful in discussing the relative merits of the datasets used, the availability of additional or alternative datasets and how key improvements in the model could be made. Horizontal integration across the different agencies, disciplines and job classifications, with regard to consensus on the data used in the model and with regard to using the model, posed surprisingly few problems in the evaluation sessions. It is likely that some operational and/or interpretational issues may arise, however, when the model is actually used in real-world decision situations. Members of the WALTER team are addressing these potential challenges through maintaining interactions with key entities via other projects, meetings, conferences and similar opportunities.

Two key challenges with regard to horizontal integration remain. The first is to provide experts with the ability to modify datasets in the model's layers. The second is to ensure that the model can be adapted to other sites in the region. Work on both of these challenges is anticipated to continue, as funding availability permits.

The WALTER website is an essential component of the project, not only because it provides the platform for access to the model, but also because the website provides an array of ancillary information ranging from descriptions of the main policies implicated in fire management decision making to animated NDVI maps, fire history information and statistical analyses of climate conditions, as well as links to many other websites. Considerable effort has gone into design of the website, particularly with regard to assuring its accessibility and usability for a wide range of individuals. The site is set up to allow different levels of access, from completely open to experts only, and provides a clear path through which users can explore without losing connection with main entry pages. Construction of the WALTER website posed its own set of challenges, not the least of which was employment of a high-caliber team of web designers and content specialists with expertise in building user-friendly and architecturally sound systems. The development process was both time- and labor-intensive and required considerable effort on the part of all researchers on the project to assure that the entire work load did not fall on the team's shoulders at the very end of the project. The effort required building sustained interdisciplinary relationships among the project's researchers, student research assistants and staff members through frequent meetings and other forms of interaction to keep abreast of the state of development of the various components of the project and to harmonize the different elements in a coherent manner.

Vertical Integration

As noted above, the architecture of FCS-1 combines GIS layers embedded in the two sub-models, the fire hazard sub-model and the values at risk sub-model. Each of these sub-models is constructed from a series of data layers representing the critical factors influencing fire risk in the forests of the Southwest US, as well as similar areas elsewhere in the West (see Figure 3.3 for an example of how FCS-1 displays individual GIS layers). Vertical integration involved:

1. assuring that all data were developed at a common grid scale of one kilometer square;
2. using common databases for underlying information such as vegetation type;
3. producing an evaluation of the quality of the data contained in each GIS layer; and
4. assuring that each layer was compatible with use of AHP.

The availability of metadata for each layer was also assured wherever possible. All of these efforts have insured that the model is as transparent as possible, and thus maximally useable and useful, especially to the experts who will be responsible for maintaining the data layers for their geographical area. An additional and absolutely crucial factor in the vertical integration process was the considerable effort expended by research team members to work together across disciplines and areas of expertise to assure that the data each team was creating would harmonize with that produced by the other teams. Another critical quality-control process in the vertical integration phase involved bringing representatives of both the expert and general public user communities in each study area to evaluate the model, its individual layers, and the data and processes used to create each layer. Participants also participated in an AHP exercise to understand how to create their own model outputs, and to assess the specific output arising from their group's AHP activity.

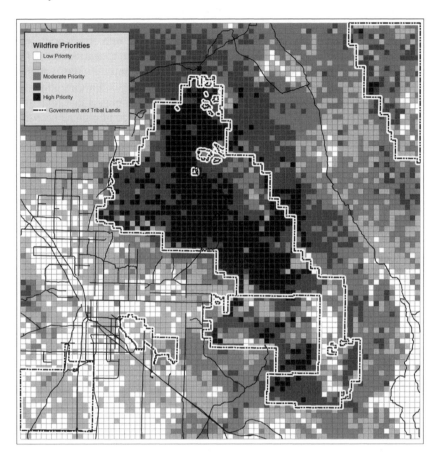

Figure 3.3   Example of FCS-1 data layer: perceived values layer

FCS-1 AS A POTENTIAL COPING MECHANISM IN BUILDING RESILIENCE TO CLIMATE IMPACTS

As recent drought conditions in the Southwest US have shown, climate impacts on the forests of the Southwest can be huge. Likewise, research that looks deep into the past shows strong links between large regional fire years and seasonal to interannual climate patterns, notably variability in precipitation associated with ENSO. Research undertaken in building FCS-1 also shows that the combination of low precipitation levels during the winter preceding fire season and high temperatures during the following spring correlate with a higher probability of occurrence of large fires in the Southwest. Through incorporation of this information into the model, and through requiring users to choose from a range of climate scenarios to run the model, FCS-1 affords a unique opportunity to think through the potential implications of climatic variability on wildland fire regimes. Further, the need to use AHP to assign weights to the model's components means that each user must think through the relationship between his/her choice of climate scenario and the impacts of that scenario in terms of fire probability and of values at risk. The results produced by the model likewise provide considerable information useful for carrying out strategic planning for time periods ranging from a month out to multiple years in the future. The ability to experiment with different climate scenarios and different weighting schemes affords opportunities to ask 'what if' questions, an exercise that can be very valuable in developing the kinds of flexibility in planning and decision making needed to cope with different possible conditions. Combined with the additional information provided on the WALTER website, the system presents valuable opportunities for enhancing resilience to climate impacts among experts and lay people alike who are concerned about the potential for and impacts of large wildland fires.

CONCLUSIONS

Wildland fire will continue to be a serious problem into the indefinite future. Getting a grip on the problem requires attention to the ways in which climate variability and change, ecological variability and change, and societal dynamics and values interact with each other and with fire. FCS-1 is a first-generation fire–climate–society integrated GIS decision support tool that brings together these disparate factors in a manner that allows assessment of conditions and alternatives at the scale of individual mountain ranges. The model maximizes possibilities for user influence over the weights assigned to the various layers, and allows for multiple experiments with different combinations of climate scenarios and weighting schemes. Subsequent versions of the model will include introduction of dynamic vegetation and climate variables, as well as web-based capacity to carry out the survey and cognitive mapping exercise used to develop the personal values layer of the model. In addition, plans call for introducing the model to other areas of the Southwest that are at risk from wildland fire.

The experience of building FCS-1 and introducing it to users in the Southwest through the model evaluation workshops held in each of the study areas has provided opportunities to acquaint key potential users with the system, to identify additional data sources to improve data quality in several of the layers, and to obtain advice on how to improve the model in other ways. The workshops also highlighted the value of the model, when used in group settings, in fostering dialogues that lead to better understanding of the sources of differences, and sometimes disagreement, between individuals and organizational units. Instances where participants in evaluation sessions asked to go back and look at certain model layers to compare weightings as they affect fire hazard and risk revealed the extent to which the flexibility of the model could be turned to creative uses in group dialogues.

FCS-1 represents one of a new generation of fire management models that are breaking new ground in terms of integration across areas of expertise and involvement of constituencies in all phases of research and development. FCS-1 arose from suggestions raised at a climate–fire workshop organized by researchers at the University of Arizona in 2000 and, from the beginning, involved interactions with individuals who would be expected to use the model (or its outputs) when carrying out strategic planning for fire management. The process offered a valuable opportunity to bridge the science–society gap in a manner that opens participation to a wider array of people who are interested in strategic planning aimed at coping with the uncertainties of managing fire in a context of uncertainty about environmental and societal change.

NOTE

1.   Information on the WALTER initiative and the FCS-1 model is available at http://walter.arizona.edu.

REFERENCES

Baisan, C.H. and T.W. Swetnam (1990), 'Fire history on a desert mountain range: Rincon Mountain Wilderness, Arizona, US', *Canadian Journal of Forest Research,* **20**: 1559–69.

Baker, W.L. and K.F. Kipfmueller (2001), 'Spatial ecology of pre-Euro-American fires in a southern Rocky Mountain subalpine forest landscape', *Professional Geographer*, **53**(2): 248–62.

Bonczek, R.H., C.W. Holsapple and A.B. Whinston (1981), *Foundations of Decision Support Systems*, New York: Academic Press.

Brown, D.P. and A.C. Comrie (2002), 'Spatial modeling of winter temperature and precipitation in Arizona and New Mexico, US', *Climate Research*, **22**: 115–28.

Brown, J.K. (1985), 'Fire effects and applications of prescribed fire in aspen', In K. Sanders and J. Durham (eds), Proceedings: Rangeland Fire Effects – A Symposium, 27–29 November, 1984. Boise, ID: United States Department of Interior Bureau of Land Management, pp. 38–47.

Comrie, A.C. and B. Broyles (2002), 'Variability and spatial modeling of fine-scale precipitation data for the Sonoran Desert of Southwest Arizona', *Journal of Arid Environments*, **50**: 573–92.

Conard, S.G., T. Hartzell, M.W. Hilbruner and G.T. Zimmerman (2001), 'Changing fuel management strategies: the challenge of meeting new information and analysis needs', *International Journal of Wildland Fire*, **10** (3-4) 267–75.

Courtney, J.F. (2001), 'Decision making and knowledge management in inquiring organizations: toward a new decision making paradigm for DSS', *Decision Support Systems*, **31**: 17–38.

Covington, W.W. and M.M. Moore (1994a), 'Southwestern ponderosa pine forest structure and resource conditions: changes since Euro-American settlement', *Journal of Forestry*, **92**(1): 39–47.

Covington, W.W. and M.M. Moore (1994b), 'Postsettlement changes in fire regimes ecological restoration of old-growth ponderosa pine forests', *Journal of Sustainable Forestry*, **2**: 153–81.

Covington, W.W., P.Z. Fulé, S.C. Hart and R.P. Weaver (2001), 'Modeling ecological restoration effects on ponderosa pine forest structure', *Restoration Ecology*, **9**(4) 421–31.

Crimmins, M.A. and A.C. Comrie (2004), 'Interactions between antecedent climate and wildfire variability across southeast Arizona', *International Journal of Wildland Fire*, **13**: 455–66.

Grissino-Mayer, H.D. and T.W. Swetnam (2000), 'Century-scale climate forcing of fire regimes in the American Southwest', *Climate Research,* **21**: 219–238.

Hartmann, H.C., R. Bales and S. Sorooshian (2002), 'Weather, climate, and hydrologic forecasting for the US Southwest: a survey', Climate Research, **21**(3): 239–58.

Henry, M. and S.R. Yool (2002), 'Characterizing fire-related spatial patterns in the Arizona Sky Islands using Landsat TM Data', Photogrammetric Engineering and Remote Sensing, **68**(10): 1011–19.

Keane, R.E., R. Burgan and J. vanWagtendonk (2001), 'Mapping wildland fuels for fire management across multiple scales: integrating remote sensing, GIS and biophysical modeling', *International Journal of Wildland Fire*, **10**(3-4): 301–19.

Kolb, T.E., M.R. Wagner and W.W. Covington (1994), 'Concepts of ecosystem health: utilitarian and ecosystem perspectives', *Journal of Forestry*, **92**(7): 10–15.

McHugh, C.W., T.E. Kolb and J.L. Wilson (2003), 'Bark beetle attacks on ponderosa pine following fire in northern Arizona', *Environmental Entomology*, **32**(3): 510–22.

Moran, M.S., T.R. Clarke, Y. Inoue and A. Vidal (1994), 'Estimating crop water deficit using the relation between surface–air temperature and spectral vegetation index', *Remote Sensing of the Environment*, **49**: 246–63.

Morgan, P., C.C. Hardy, T.W. Swetnam, M.G. Rollins and D.G. Long (2001), 'Spatial data for national fire planning and fuel management', *International Journal of Wildland Fire*, **10**(3-4) 353–72.

National Interagency Fire Center (NIFC) (2004a), *National Wildland Fire Outlook*, Predictive Services Group, National Interagency Fire Center, Boise, Idaho, Issued 6 February 2004.

National Interagency Fire Center (NIFC) (2004b), *Wildland Fire Statistics*, National Interagency Fire Center, Boise, Idaho, http://www.nifc.gov/stats/wildlandfirestats.html, Accessed 23 March 2004.

Nemani, R., L. Pierce, S. Running and S. Goward (1993), Developing satellite-derived estimates of surface moisture status', *Journal of Applied Meteorology*, **28**: 276–84.

Nicholson, C.R., A.M. Starfield, G.P. Kofinas and J.A. Kruse (2002), 'Ten heuristics for interdisciplinary modeling projects', *Ecosystems*, **5**: 376–84.

Overpeck, J.T., D. Rind and R. Goldberg (1990), 'Climate-induced changes in forest disturbance and vegetation', *Nature*, **343**: 51–53.

Pyne, S.J. (1982), *Fire in America: A Cultural History of Wildland and Rural Fire*, Princeton, NJ: Princeton University Press.

Pyne, S.J. (2001), *Fire: A Brief History*, Seattle: University of Washington Press.

Pyne, S.J., P.L. Andrews and R.D. Laven (1996), *Introduction to Wildland Fire*, 2nd edn New York: John Wiley and Sons, Inc.

Reed, B.C., J.F. Brown, D. VanderZee, T.L. Loveland, J.W. Merchant and D.O. Ohlen (1994), 'Measuring phenological variability from satellite imagery', *Journal of Vegetation Science*, **5**: 703–14.

Running, S., R. Nemani, D. Peterson, L. Band, D. Potts, L. Pierce and M. Spanner (1989), 'Mapping regional forest evapotranspiration and photosynthesis by coupling satellite data with ecosystem simulation', *Ecology* **70**(4): 1090–1101.

Saaty, T.L. (1980), *The Analytic Hierarchy Process*, New York: McGraw-Hill.

Saaty, T.L. (1990), *Multicriteria Decision Making: The Analytic Hierarchy Process*, Pittsburgh: RWS Publications.

Schmoldt, D.L. and D.L. Peterson (2000), 'Analytical group decision making in natural resources: methodology and application', *Forest Science*, **46**(1): 62–75.

Sheppard, P.R., A.C. Comrie, G.D. Packin, K. Angersbach and M.K. Hughes (2002), 'The climate of the Southwest', *Climate Research*, **21**: 219–38.

Simard, A.J., D.A. Haines and W.A. Main (1985), 'Relations between El Niño Southern Oscillation anomalies and wildland fire activity in the United States', *Agricultural and Forest Meteorology*, **36**: 93–104.

Sprigg, W. and T. Hinckley (eds) (2000), *Climate Change in the Southwest*, ISPE, University of Arizona, Tucson.

Swetnam, T.W. and J.L. Betancourt (1990), 'Fire–southern oscillation relations in the southwestern United States', *Science*, **249**: 1017–20.

US Congress (2003), *Healthy Forests Restoration Act of 2003: HR 1904*, 108th Congress of the United States of America, 7 January 2003.

White House (2002), 'Healthy forests: an initiative for wildfire prevention and stronger communities', http://www.whitehouse.gov/infocus/ healthyforests/ Healthy_Forests_v2.pdf, Accessed March 2004.

White, S.M. (2004), 'Bridging the worlds of fire managers and researchers: lessons and opportunities from the wildland fire workshops', USDA Forest Service Pacific Northwest Research Station: Portland, Oregon. General Technical Report PNW-GTR-599, March 2004, 41pp.

Yool, S.R. (2000), 'Enhancing fire scar anomalies in AVHRR NDVI time-series data', *GeoCarto International*, **16**(1): 5–12.

4 An Integrated Assessment of Impacts of Predicted Climate Change on the Mackinaw River Basin

K. Donaghy, W. Eheart, E. Herricks and B. Orland

INTRODUCTION

There is overwhelming evidence that global climate change is occurring. USAID (1998) reports that since 1860, atmospheric CO_2 concentration has increased by 30 percent, while average global temperature has risen by 0.5°C. By 2100, the average global temperature is expected to increase by 1 to 3.5°C. Greater increases are anticipated in areas closer to the poles, and smaller increases in areas closer to the equator. Manifestations of climate change will not be limited to temperatures. Also varying will be the degree and timing of regional climate extremes, such as the length of growing seasons and soil moisture. Changes in seasonal pattern and variability of precipitation will lead to either wetter or dryer environments (because evaporation will more than compensate for increased precipitation) and increased danger of floods and high winds. The frequency of extreme weather events will increase and, by 2100, the sea level will rise by nearly 50 cm. When combined with storm surges and tides, this rise will lead to incursions of seawater into coastal and freshwater areas and significant erosion. These changes will affect not only natural systems, but also economic and social systems, disrupting agricultural production, manufacturing and transportation and increasing vulnerability of human health.

Although climate change is occurring globally, it is experienced locally (or regionally), where people live and work. In the case of the Midwestern United States, significant impacts have already been experienced. Lettenmaier et al. (1994) (cited in IPCC 1998) have identified increasing trends both in precipitation during the months of September to December and in streamflow during the months of November to April over the period 1948–88. This is particularly significant because even small changes in mean annual rainfall have greater impacts on water resources than changes in mean hydrological conditions (IPCC 1998).

In the state of Illinois, extreme weather events have been experienced increasingly over the last twenty years. These events include the severe drought of 1988, the flood of 1993, the 1995 heatwave, a severe rainstorm in Chicago in 1996, the 1997–1998 El Niño weather pattern, the 1999 heatwave and the 1999 windstorm in Bloomington. These events have imposed heavy costs upon the state. The flood of 1993 alone imposed over $8.9 billion of damages to agriculture, $1.9 billion to transportation, plus soil erosion, water quality degradation, groundwater contamination and the spread of exotic species such as zebra mussels. Human costs included 52 deaths, displacement of 94 000 people, months of stress, anxiety and insecurity, and disruptions to commuting. The 1995 heatwave cost upwards of 250 Chicago residents their lives and a single severe rainstorm in Chicago in 1996 imposed costs on the city and state close to $600 million. In view of these damages, the Illinois state government has established a task force to prepare mitigation plans for the effects of extreme climate events.

While policy discussions are continuing at the international level over how or if the terms of the Kyoto Protocol will be implemented by nations (see for example Carraro and Goria 1999), at the sub-national level communities and regions must respond and plan to respond to the problems brought on by the effects of global climate change. These problems include not only droughts, floods and periods of temperature extremes, but also the increased incidence of windstorms, severe thunderstorms and tornadoes. Impacts will be felt in damage to infrastructure and housing, stress on public utilities and emergency services, fluctuations in pest populations, damage to crops and loss of goods in shipment or storage.

The 'problem' climate change poses at the regional level is one of understanding adequately and responding appropriately. If with Friedmann (1987) we view regional planning as the 'management of change in territorially organized systems', then coping with climate change entails coming to terms with regional impacts; getting regional stakeholders to recognize and appreciate the nature of these impacts and their short-term and long-term implications; envisioning and getting stakeholders to consider coordinated response strategies (comprehensive planning at the regional level); plotting out alternative transition paths to sustainable settlement and transportation patterns, building technologies and so on, with their associated costs and benefits; and facilitating public deliberation and political will formation. While the problem is formidable, it does not differ from other planning problems in certain crucial respects. Because of the interdependence of outcomes (system effects), actions are not separable. Because action increments (technology and infrastructure changes) often must be undertaken in indivisible units, continuous marginal adjustments are neither efficient nor possible. Because previous states cannot be returned to without cost and actions are often irreversible, history and dynamics matter. And because, as discrepancies in global circulation model forecasts suggest, more than one future is possible and foresight is imperfect, uncertainty cannot be eliminated.[1]

Description of the Research

The research discussed in this chapter responds to this problem by carrying out an integrated assessment of multiple-sector impacts on a Midwestern US watershed, produced by predicted changes in climate. The research used historical data, models, and standard and innovative analysis tools in conducting this assessment and was guided by early stakeholder input. The impact assessment focused on locations in the Mackinaw River watershed in Illinois. The specific objectives were:

1. develop sector-specific responses to climate change;
2. identify relationships between, and among, sectors at each site, and among all sites;
3. apply the impact analysis paradigm to identify and quantify local impacts produced by climate change;
4. identify mechanisms that produce an adaptive response to climate change while developing sector/system resilience to climate change impact; and
5. integrate project results with a web-based decision support interface available at the University of Illinois and Pennsylvania State University.

Responsibilities of the research team members over the course of the project were as follows. Brian Orland led development of the web-based interface for the project. Wayland Eheart was primarily responsible for SWAT modeling activities, and provided both flow and crop production data for research by J. Edwin Herricks and Kieran Donaghy. Donaghy assumed the responsibility for econometric modeling, which was intended to integrate with other parts of the project, particularly Eheart's predictions of the effect of climate change on agricultural systems. Herricks was responsible primarily for the natural resources assessment activities with a focus on the translation of flow variability issues into information on aquatic ecosystem impacts, and consideration of small-scale alterations in habitat that might be produced by extreme events. The general plan for the integrated effort was to use the stakeholder input as the foundation for integration, while conducting focused research in areas of individual responsibility to develop an analysis and modeling structure for incorporation into the DSS.

The research project was initiated as an integrated study to assess climate change impacts on a regional scale. The project was designed to identify and quantify the consequences of climate variability and change on human and natural systems in the Mackinaw River Basin in Illinois. The Mackinaw River watershed of approximately 3100 km^2 (1200 mi^2) provides an ideal arena for both identifying local impacts and integrating the assessment across sectors. The Mackinaw River corridor is one of the highest quality ecosystems in Illinois with water quality rated as excellent but threatened. Identified threats include agriculture and the influence of urban centers. The Mackinaw has been the focus of recent, integrated management activity, which includes major involvement of the Nature Conservancy, several Illinois agencies and local watershed groups. The focus of the proposed research was the sensitivity and vulnerability of local human and natural systems, and adaptive mechanisms that operate in response to impact threats.

In our working approach, we identified *system* categories and associated *sectors* for the Mackinaw watershed. Human environment systems have industrial, agricultural, municipal and regulatory activities defining sectors. The physical environment system has water resources and landscape features (for example topography, roads and drainage systems) as the principal sectors. Living systems in the environment have aquatic ecosystems, terrestrial ecosystems and human health as the principal sectors.

We identified vertical and horizontal relationships for systems and sectors. An initial vertical and horizontal integration used a simple matrix concept to assist in accounting for interactions between and among sectors. In this accounting we identified sectors within systems. In each sector, a number of *elements* were defined. An element is a characteristic or condition of the sector that will respond to climate change. For example, in the agricultural sector, crop selection, crop rotation schedule, choice of hybrid, or application of fertilizers and pesticides are elements that may be affected by climate change. In this working approach there are multiple elements in each sector, some directly affected by climate change, others only indirectly affected. Our approach also recognized that the potential interaction between and among sectors is actually based on the interactions between, and among, elements. We also argued that integration between, and among, elements may change based on the type of impact (ecosystem degradation, social discord, economic loss and so on) that is the focus of the integration. In summary, our horizontal integration occurs between, and among, sector elements, and our vertical integration is based on specific climate system changes and layers of cross cutting analysis that are defined by impact type. This working approach provided an organized method to address the inherent complexity of climate change impacts, while providing a framework for the application of analysis and modeling tools that will develop accurate and sensitive predictions of local impact due to climate change.

The impact assessment paradigm selected for this research was implemented in the following manner. In a local setting in the Mackinaw River watershed, systems and sectors that operate in, or have influence on, the local setting were identified. For identified sectors, an 'elemental' analysis was performed to identify how each element responds to climate change phenomena either directly or indirectly. Indirect effect analysis used an empirical analysis and a quantitative approach. We also planned to identify chains of effect that are known and quantified, or are supported by existing models. The sector specific analysis was intended to evaluate the response of each element to five categories of climate change phenomena:

1. a long-term increase in average temperature and CO_2;
2. altered timing of annual change in local weather conditions (for example shifting dates of frost free conditions, growing season start or length and so on);
3. a change in weather conditions (for example shift from soaking rain to convective storms);
4. increased magnitude of extreme weather events (such as floods or droughts); and
5. altered frequency of extreme or common weather events.

The intent of the research was to develop elemental response spectra that are site specific. Responses would then be assessed against impact criteria to complete the impact assessment. In this research there is also vertical integration initially based on the five identified climate change phenomena and 'layers' of related elements that are defined by impact type. For example a social or economic impact may involve specific elements in each sector. In this research, these layers will be defined by the *impact types:* socioeconomic, environmental and water resources.

Additional layers of this cross cutting analysis may be defined by a designated sector. For example, agricultural interests may dominate at a location. If so, that sector may dominate an analysis and actually produce impact in other sectors. Assessing a sector-driven impact will benefit from collaborator involvement in this research.

Key in this research approach are the availability of site specific information; knowledge of element/sector/system aggregation possibilities; and availability of analysis methods and modeling tools to integrate across sectors and in specified layers. We expected that site specific information and important sector specific insights would be provided by the extensive historical data resources available, recent condition and trend compilations for the area, identified strategies for dealing with climate change (Illinois Department of Energy and Natural Resources 1994)[2] and the input from local collaborators. Full competence in all aggregation possibilities and all of the analysis methods and models was considered to be beyond the capacity of any small research team so we emphasized the use of stakeholder input to assist in defining system and sector relationships.

The Role of Stakeholder Workshops in Shaping the Research

We proposed and held a series of workshops/study sessions with stakeholders, where we used the results of our recently completed study on the adjacent Sangamon River to identify a range of 'change/impacts' associated with climate change. We shared this information with local groups with the intent of meeting our citizen data collection and citizen validation needs early in the project. We conducted approximately 80 interviews with various stakeholders, and held four focus group workshops whose objective was to encourage interaction among individuals who had been interviewed to develop a sector 'interaction' analysis from the stakeholders. The interviews and workshops were very valuable. The workshops were held at locations throughout the basin and drew participants from sub-regions. The following provides a summary of workshop results.

There was no agreement on what was the most important weather extreme, but there was general agreement that events that had a long effect time (such as icestorms and tornadoes) were most significant. There was a general feeling that in the climate change scenarios presented/understood that sectors would be able to adapt. A consistent theme was that it would be possible to work better and smarter with equipment – new technology would solve problems. An important finding concerning the human dimension of climate change is that volunteerism is viewed as an asset in small communities, which may allow them to cope with change or significant events better than larger communities. Another common theme was that adaptation produced by upgrading infrastructure, such as placing utilities underground, would address significant impacts of climate change. There was a general sense that quality of life issues would drive investment to meet changing needs produced by climate change. It was recognized that changing climate could have an effect on crops, cropping and general agricultural practices. Adaptation time for changing agricultural practices was estimated at between 10 and 20 years, which would be the critical time interval when considering adaptive processes. Economic effects are likely to be significant if the frequency of extreme events increases. For example, a local utility may not have sufficient cash reserves to deal with multiple events that require significant capital for repairs. It was recognized that social change will accompany climate change and is associated with a wide range of issues that may vary from individual response, through different expectations of communities, to more 'global' change produced by economic alterations in the system.

The workshops were helpful to the research team in identifying a number of questions, which, among others, included:

1. When does a challenge become a problem/impact?
   It was very apparent from the interviews and the stakeholder workshops that one individual's, or one group's, critical issue was of minimal concern to others. One factor that did stand out was the time interval of event versus time interval of response. The concern identified was that the expectations of the public will change faster than climate change will occur and the mismatch in the system will slow both response and recognition.

2. What is the climate change constituency?
   We came to realize that there was a significant resistance to the use of terms 'global warming' and there was a strong sentiment that science was not fully supportive of a climate change scenario. Although some individuals were not convinced of the need for action, others recognized the potential consequences of climate change and were willing to consider needed changes in policy or operations. Although there was no real constituency for climate change preparedness, there did seem to be a strong constituency for dealing with extreme events. There was enough 'memory' of severe weather to support the position that some items are worth spending money on even though costs can not be justified.

The implementation of the stakeholder meetings at the outset of the project altered the proposed project in several major ways. First, it demanded that we review our initial development of system/sector/element characterizations and provided a basis for consideration of new vertical and horizontal integration opportunities. The early stakeholder involvement also provided a basis for a realistic assessment of stakeholder interests and the possibilities of future involvement. We also came to understand that individual stakeholders had individual interests and that any decision support system development would require careful consideration of individual needs balanced against both system capacity and the need for general applications.

One outcome from the workshops was the use of advanced technology to provide access to workshop discussions in a web-based environment, which was made available on the web via the decision support system. Those recordings were then used to develop a text summary of the workshops, which was made available on the web to project results and the decision support system. It is possible to search this summary text for key words and then retrieve full sections of recorded dialog to provide a complete context of the discussions. This part of the project was initially made available to the project team only, and then was added as an element of the decision support system.

The Role of Web-based Decision Support System (DSS) in the Research

Following the stakeholder meetings, the research elements of the project were evaluated and the investigators began to develop specific plans to meet project requirements with the focus provided by stakeholders. The project interface, which will be referred to as the DSS, was intended to be a decision support tool that integrated data, models and other elements to convey information about climate change. A particular challenge faced was the development of a 'visualization' tool that effectively supported the entry of the range of interests represented in the stakeholders into the DSS process, while also integrating the results of the natural resources, hydrology, water quality and economic analysis and modeling in the system.

The structure and appearance of the DSS evolved through the project. Two of the primary elements of the DSS were the workshop transcript analysis tool previously described, and a system to provide summaries of historical weather to evaluate climate change potential. This historical weather system, whose development was facilitated by the assistance of Dr Kenneth Kunkel of the Illinois State Water Survey, included data for 100 years of observations for 40 stations. We developed an interface tool permitting the data to be accessed, queried and plotted in a user-friendly fashion. We have collected historical records and documents for development of a database of qualitative information. This product was available early in the project and provided a means to convey potential climate change conditions to stakeholders. For example, it was possible to identify extreme events from the past, or identify years with unusually hot or cold temperatures to illustrate to stakeholders the possible consequence of climate change. (see Table 4.1.) This inquiry system also provided a valuable asset to the research team, who used historical climate data to assess model results as part of the calibration/validation process.

Plan of the Chapter

In the remaining sections of this chapter we discuss examples of specific research undertakings concerned with assessing regional impacts of climate change on water resources, agricultural systems, the economy and fish populations. We also discuss how these projects are related and what has been accomplished in development of the project DSS. The chapter closes with a brief assessment of the accomplishments of the project overall.

Table 4.1   Results of query of historical climate database

Average precipitation in inches (Precipitation)	5.5
Maximum daily high temperature in °F (Tmax)	89.7
Minimum daily high temperature in °F (Tmin)	64.2
Average daily high temperature in °F (Tavg)	77
Total snowfall in inches (Snow)	0
The highest daily maximum temperature in °F (Thigh)	105.5
The lowest daily maximum temperature in °F (Tlow)	43.5
Number of days with measurable precipitation (Prec>0)	17.5
Number of days with precipitation greater than 0.1 inches (Prec>0.1)	9.5
Number of days with precipitation greater than 0.5 inches (Prec>0.5)	4
Number of days with precipitation greater than 1 inches (Prec>1)	2
Number of days with daily maximum temperature greater than 90 °F (T>90)	51
Number of days with daily maximum temperature less than 32°F (T<32)	0
Number of days with daily minimum temperature less than 0°F (T<0)	0
Number of days with measurable snowfall (Snow>0)	0
Number of days with snowfall greater than 1 inch (Snow>1)	0
Number of days with snowdepth on ground greater than 1 inch (Dep>1)	0
Number of days with daily maximum temperature greater than 95°F (T>95)	27
Number of days with daily maximum temperature greater than 100°F (T>100)	10.5

Source: http://www.rehearsal.uiuc.edu/epa/web/models/result.asp (Climatic data shown for the Aledo station for the summer season, and temperature variance 3–4; two years of occurrence, 1901 and 1988.)

WATER RESOURCES AND AGRICULTURE MODELING

Agriculture is one of the sectors most sensitive to climate change and climate change is expected to have profound influences on the Midwestern agricultural economy and regional water resources. Much effort has been put into investigating the consequences of climate change on the Midwest. These studies employed various methods, were conducted at different scales and were based on varying scenarios of climate change projections. Many studies focused on climate change effects on the productivity of customary crops with economic importance. A study cited in the report produced by National Assessment Synthesis Team (NAST) projected a favorable future for crop yields in the Midwestern region with an exception in some locations in the southern portions of Indiana and Illinois where the corn yields could decrease by 10–20 percent (NAST 2000b). Other recent examples of this kind of study include Brown and Rosenberg (1997, 1999); Brumbelow and Georgakakos (2001); Southworth et al. (2000, 2002); and Izaurralde et al.(2003).

Concerns about water resources in an agricultural region are associated with the water demands of irrigation. In addition to its direct effects, climate change can also affect regional water resources indirectly through altering the irrigation needs. For example, Bowman and Collins (1987) examined the effects of irrigation on Illinois groundwater resources under drought conditions. In the recent climate change impact studies, Brumbelow and Georgakakos (2001) and Izaurralde et al. (2003) included irrigation need shifts in their assessments.

This section presents a discussion of a modeling exercise intended to help assess the impact of climate change on agricultural productivity and water resources vulnerability of the study basin. The modeling techniques employed permit investigation of the linkages between them through the potential irrigation demand shifts due to the climate change.

Generally in our studies we envisioned a future climate with more frequent droughts on the basis of outcomes of the well-known General Circulation Models (GCMs). As a traditional rain-fed agricultural region, irrigation is sporadic in the Midwest. The threats of droughts arising from climate change will potentially motivate farmers to introduce irrigation to meet the water requirement of crop growth and maintain high and stable yields. In our assessment the vulnerability of water resources is characterized by the low flow frequency of streams. The low flow days are a period of critical time to the aquatic ecosystem and the capacity of water bodies receiving pollutants. In the Midwest, summer is the primary growing season of crops and it is during this period that streams tend to have lowest stream flow. The irrigation induced by climate change, if any, may change the low flow frequency of the stream.

An important way our study differs from many other studies is that we introduce an economic analysis of irrigation profitability on the basis of agricultural productivity and hydrological simulation. Such an analysis can reveal how well the agricultural economy in the watershed would adapt to the climate change through irrigation and promote deeper understanding of the adaptive activities of irrigation and its effects on water resources.

The discussion in this section is divided into three parts. In what follows the methods used in the modeling exercises are described. Results and conclusions are presented in the final section.

Methods

The Mackinaw River basin is located in central Illinois (USGS Cataloging unit: 07130004). The Mackinaw River is a tributary of the Illinois River and flows west before it joins the main stream. The Mackinaw watershed drains

3000 km², has a mean elevation of 220 m above sea level and a continental climate. The Mackinaw river basin is typical of agricultural watersheds in the Midwest with varying soil types and dominant agricultural land use. Agricultural land use accounts for more than 95 percent of the total area of the watershed. Currently most of the area of the watershed is not heavily irrigated except a small fraction of the land near the outlet of the river to the Illinois river, which belongs to Havana Lowlands, a major irrigation center in Illinois state (Bowman and Kimpel 1991).

A general description of the simulation framework
In order to evaluate the impacts of climate change on the basin, scenarios representing various climatic patterns, dominant crop types and irrigation practices are designed. Simulation exercises are conducted with a model for these scenarios. By interpreting simulation results, the potential consequences of climate change on agricultural productivity, irrigation needs and water resources vulnerability of the study watershed can be determined.

The model used in the simulation exercise is the Soil and Water Assessment Tool (SWAT). It is a physically-based watershed scale model developed for watershed hydrology simulation and the evaluation of the long-run effects of land management practices (Arnold et al. 1998). The model has the capability to simulate the water movements, vegetation growth and a range of agriculture operations including irrigation. SWAT provides an integrated simulation environment to investigate the climate change impacts on agricultural productivity and watershed hydrology.

The simulation modeling exercise starts with watershed delineation. The entire watershed is delineated into sub-basins. Sub-basin boundaries are defined by watershed surface topography so that water within a sub-basin flows to the outlet of the sub-basin. To increase the accuracy of the simulation, a sub-basin may further consist of several Hydrologic Response Units (HRUs) which allows SWAT to account for diversity within a sub-basin in land use, soil type and management practices. HRUs are partitions with unique combinations of these attributes.

SWAT2000, the version of the model we used to conduct the simulations, has been incorporated into BASINS (Better Assessment Science Integrating Point and Non-point Sources) version 3.0. BASINS is an analytical software package developed by the US Environmental Protection Agency Office of Water operating within the Arcview Geographic Information System (GIS). BASINS provides GIS-based watershed delineation tools. The employment of the GIS technique facilitates the delineation and increases the accuracy. With the aid of the BASINS delineation tool, watershed delineation for SWAT modeling can be completed automatically with desired sensitivity. The data required for the delineation are distributed along with the BASINS software package. In our modeling exercises, the entire Mackinaw River Basin is portioned into 19 sub-basins, 73 HRUs (Figure 4.1).

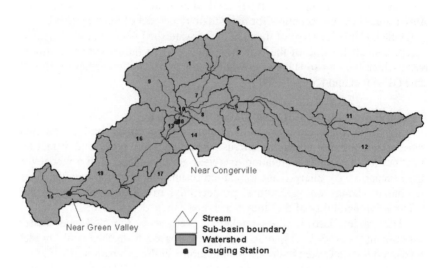

Figure 4.1   Delineation of the Mackinaw River Basin

The processes simulation of SWAT is conducted on a daily basis and driven by meteorological inputs. Daily temperature, precipitation, solar radiation, wind speed and relative humidity data are required. By modifying these input data, various climate patterns can be represented.

Combining the SWAT outputs of crop yields, price information in agriculture products market and the costs information of crop productions (including the irrigation costs, if any), the climate change impacts on the regional agricultural economy can be determined. Moreover, the economic profitability and the likelihood of the occurrence of irrigation can be examined by differencing the profits in irrigation and no irrigation cases.

Two US Geological Survey (USGS) gauging stations, one near Congerville (station number 5567500) and the other near Green Valley (station number 5568000) are chosen as reference stations. The low flow standards at the two reference gauges, here assumed to be the $_7Q_{10}$, are 0.033 m^3/s, 0.714 m^3/s respectively (Singh et al. 1988). The locations of these reference gauging stations are also labeled in Figure 4.1. Following the general description of the model exercise, more details are given below.

Climate patterns
Two climate patterns, referred to as Illinois (IL) climate pattern and Oklahoma (OK) pattern, are included in our scenarios of interest. The former

climate pattern represents the current climate in central Illinois and the latter one presents the climate of Illinois after change. The meanings of the Oklahoma climate pattern are explained below.

While 30-year historical records of temperature and precipitation during the baseline period 1963–1992 at the local weather station near the Mackinaw River Basin are used to represent the current climate in Illinois, the formulation daily meteorological input data series representing future Illinois climate is based on the outcomes of GCMs. In their report, the National Assessment Synthesis Team primarily used two climate models developed at the Canadian Climate Center and the Hadley Center in the UK. They claimed that 'these models were the best fits to a list of criteria developed for this assessment' (NAST 2000a).

Although the two climate models have similar assumptions about the future greenhouse gas emissions, there exist some differences between the climate change projections produced by the two models. Generally, the Canadian model shows a picture with a more arid climate. In our studies, the outcome of the Canadian model is used.

Given their resolutions, the outcomes of GCMs can not be directly used in our study at regional scale. Series of weather inputs on daily step required by SWAT have to be constructed. Instead of using some statistical downscaling techniques, in our study we chose to construct the weather input series by analog. In NAST reports, the projections of GCMs on the future climate in Illinois are provided in such an analogous way. According to the Canadian model, the summer climate of Illinois in 2030 would become the current climate in Missouri and in 2090, the current climate in Oklahoma State (Figure 4.2) (NAST 2000b). According to a further examination of the outputs of the transient runs of the Canadian model for 2090 (Liang, personal communication 2003), such analog holds even for winter months.

Based on the analog, temperature and precipitation series representing the climate of Illinois in 2090 are constructed by picking up historical records from the weather station in Oklahoma City (station name: Oklahoma City Will Rogers Wor, Location: 35°23'N/97°36'W, Elevation: 391.7 m, WMO_ID: 72353, Source: National Climatic Data Center) during the baseline period 1963–1992. The monthly mean temperature and precipitation representing the Oklahoma pattern are listed in Table 4.2.

The observed data of daily solar radiation, wind speed and relative humidity are not available from records. Such series are generated by the WXGEN weather generator (Sharpley and Williams 1990) and incorporated in SWAT according to their monthly statistics of the reference station in Oklahoma City contained in the database distributed with BASINS software package. Another climatic factor that could profoundly change the appearance of the watershed is the increased concentration of carbon dioxide (CO_2) in the atmosphere. In the two primary climate model simulations used in NAST assessment, it is assumed that the greenhouse gas emissions will result in a CO_2 concentration of just over 700 ppm (NAST 2000a). The beneficial effects of elevated CO_2 on plant growth are well-known, especially in the dry climate. Such effects decrease the stomatal conductance and the transpiration of plants, which potentially increases the crop yields and changes the watershed hydrological condition. To determine the effects of elevated CO_2 concentration, under the Oklahoma climate pattern, holding the other climatic factors constant we run the model at $CO_2 = 350$ ppm and $CO_2 = 700$ ppm.

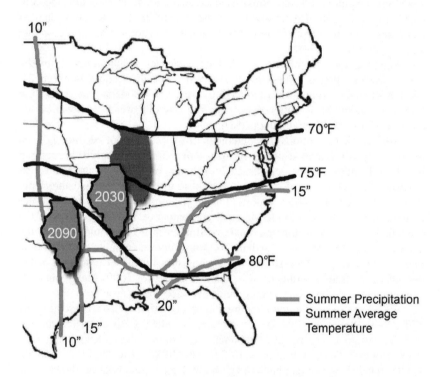

Source: Adapted from NAST 2000b, p. 55.

Figure 4.2   Climate change projection for summer Illinois by Canadian Climate Center Model

Alternative crops
Three alternative crops, corn, soybean and winter wheat double-cropped with soybean, are included in our studies. Traditionally corn and soybean are two primary crops which account for most acreage of crops in Mackinaw River Basin. The acreage of winter wheat in the two watersheds is negligible

currently. Winter wheat has to be double-cropped with soybean in order to produce profits competitive with corn and soybean. However, the short summer growing season under the current climate make the double-cropping 'too risky' (Nafziger, personal communication 2003). The warming trend of climate change will increase the length of the growing season and make the double-cropping a promising choice to farmers.

Table 4.2   Monthly mean temperature and precipitation under Illinois and Oklahoma climate patterns

	Illinois		Oklahoma	
	Precipitation (mm)	Temperature (°C)	Precipitation (mm)	Temperature (°C)
Jan.	39.2	-4.4	28.8	2.3
Feb.	38.6	-2.0	38.3	5.1
Mar.	76.9	4.7	67.6	10.3
Apr.	97.2	11.6	72.8	16.0
May.	105.7	17.3	138.4	20.2
Jun.	94.2	22.6	108.1	24.9
Jul.	100.1	24.4	65.8	27.9
Aug.	84.7	23.1	69.3	27.1
Sep.	92.6	19.5	101.2	22.8
Oct.	66.6	13.0	82.0	16.6
Nov.	76.3	5.6	51.4	9.4
Dec.	79.9	-1.5	41.6	4.0

Growing seasons and the irrigation periods of the three alternative crops we used in the simulation are shown in Table 4.3.

Table 4.3   Growing seasons and irrigation periods of alternative crops

Crops	Growing season	Irrigation period
Corn	Early May–Mid October	Early May–Early August
Soybean	Early May–Early September	Mid May–Late August
Double-cropping:		
Winter wheat	Early October–late June (including dormancy)	Early April–Late May and Early October–Middle October
Soybean	Early July–late September	Early July–early September

Climate change will affect the growing seasons and irrigation periods. However, experimental runs show that if the scheduling are roughly within the range shown in Table 4.3, no fundamental difference in our conclusion will be made.

Model performance

The simulation of phonological development of plants in SWAT is based on potential heat unit theory and controlled by parameters in the plant growth database. Those parameters account for genotype differences of various plants. At every daily step of plant growth simulation, the optimal plant development is first computed and then adjusted for temperature, moisture and nutrients stresses.

Take the computation of biomass development as an example (Neitsch et al. 2001).

$$\Delta bio_{act,i} = \Delta bio_i \cdot \gamma_{reg}$$

Where $\Delta bio_{act,i}$ is the actual increase in biomass on a given day i (kg/ha), Δbio_i is the potential increase in biomass in that given day (kg/ha), and γ_{reg} is the plant growth factor (0.0–1.0). Further,

$$\gamma_{reg} = 1 - \max(wstrs, tstrs, nstrs, pstrs)$$

where *wstrs*, *tstrs*, *nstrs* and *pstrs* denote daily water stress, temperature stress, nitrogen stress and phosphorous stress. Currently, among the four stresses simulated by SWAT, the nutrients are not limiting to crop yields (Nafzider, personal communication 2003). Hence, in our simulation we assumed that crop will not experience nutrient stresses. The nutrient stress simulation function of the model was switched off.

The model performance in crop yields simulation was validated to observed yields. The observed crop yield data are from the US Department of Agriculture (USDA) database and Illinois Farm Business Farm Management Association (FBFM). The model performs well in simulating corn and soybean yields. There are upward trends of crop yields since the 1960s that could be attributed to the advances in technology and management. The simulated corn and soybean yields are roughly at a level corresponding to that in the 1980s. During this period, the simulated and observed yields agree well. The simulated corn and soybean yield series can follow the fluctuations of historical records.

The simulated winter wheat yields do not agree well with the observed data. There are several possible explanations for this problem: first, the acreage of winter wheat in the study region is too small so that the yields from historical records are not representative; or the algorithms in SWAT cannot account for some factors that influence crop yields other than temperature and moisture, such as insects, diseases and weeds and so on. Some of these factors may affect the yields of winter wheat significantly. As for the double-crop soybean, there are no official records on its yield available. But it is estimated that the yield potential of double-crop soybean is only 50–60 percent of that obtained with time planting (Nafziger 2003). In our simulation, we made the model give reasonable estimates of annual average yields for the yields of winter wheat and double-crop soybean through calibration.

As a physically-based model, SWAT allows explicit incorporation into the modeling exercise of knowledge obtained in other studies on the fertilization effects of CO_2. In SWAT, the default ambient CO_2 is 330 ppm. Increased CO_2 first calls for an adjustment of radiation–use efficiency of the plant, which is the amount of dry biomass produced per unit intercepted solar radiation (Stockle et al. 1992) and the modification to the leaf conductance term (Easterling et al. 1992). Radiation-use efficiency of a plant, RUE $(kg/ha[MJ/m^2]^{-1})$ is defined as

$$RUE = \frac{100 \cdot CO_2}{CO_2 + \exp(r_1 - r_2 \cdot CO_2)},$$

where CO_2 is the ambient CO_2 concentration (ppm), and r_1 and r_2 are shape coefficients. Leaf conductance, g_{l,CO_2}, which is defined below, is modified to reflect CO_2 effects (m s^{-1}):

$$g_{l,CO_2} = g_l \cdot [1.4 - 0.4 \cdot (CO_2 / 330)],$$

In a recent field experimental study at the University of Illinois, the effects of nitrogen supply and CO_2 elevation on corn yields are shown to be additive. At the nitrogen supply rate of 202 kg/ha, an increase of CO_2 from 370 ppm to 550 ppm will increase the corn yield by 20 percent (Uribelarrea et al. 2003). The percentage of increase meets the estimates we got from the model simulation under the two levels of CO_2. The capability of hydrological capacity of SWAT has been validated at various sites (Bingner 1996; Santhi et al. 2001; Van Liew and Garbrecht 2003). In our studies, the model is manually calibrated for streamflow at reference stations. The emphasis of the streamflow calibration is placed on the fits of the low flow period. In general the model can capture the characteristics of the streamflow in low flow season.

Irrigation simulation
In SWAT simulation, the occurrence of an irrigation event is triggered by Root Zone Deficit (RZD). RZD is the difference between soil field capacity and the actual soil moisture. Once RZD achieves a threshold value for irrigation, irrigation will occur.

Farmers could either apply groundwater or abstract water from the stream for irrigation. Cost information of irrigation is shown in Table 4.4. For most

of the area of the Mackinaw River Basin that is not irrigated, the total
irrigation costs consist of both a fixed capital component required for
irrigation facility construction and a variable portion that varies with the
amount of water applied. Surface water irrigation is less costly but only
available in the riparian area. We define the riparian area where irrigation
could occur as that area within 1 mile of the stream and with non-zero $_7Q_{10}$ in
corresponding reaches. The area is labeled in Figure 4.3.

Table 4.4   Irrigation costs (Wright and Benson 1983)

	Annual capital costs ($/ha-yr)	Variable costs ($/ha-mm)
Surface water irrigation	109.96	0.2
Groundwater irrigation	165.93	0.27

Source: Wright and Benson 1983

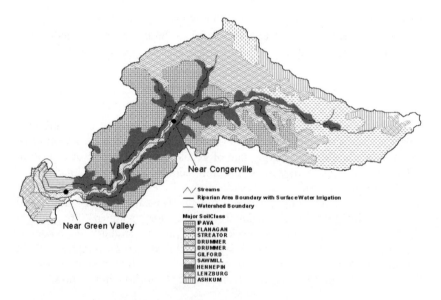

*Figure 4.3   The soil distribution and riparian area boundary in the
Mackinaw River Basin*

The irrigation is further assumed to be unregulated and the capacity of the
irrigation facility to be 40 mm/d. In an irrigation event, soil moisture will
attain to its field capacity unless daily capacity of the irrigation facility is
achieved; or in surface water irrigation the streamflow falls to zero. For

groundwater irrigation, there is no constraint on water availability. The groundwater is assumed to be taken from deep aquifers and the storage in deep aquifers is assumed to be unlimited.

The formulae for irrigation profitability calculation are given below. In the case of no irrigation, the formula is:

$$profit_{noirr}(\$/ha) = yield(ton/ha) \times c.f. \times price(\$bu/ha) - production\ costs(\$/ha)$$

whereas in the case of irrigation the formula is:

$$profit_{irr}(\$/ha) = yield(ton/ha) \times c.f. \times price(\$bu/ha) - production\ costs(\$/ha)$$
$$- irrigation\ costs(\$/ha).$$

Hence, the profits increase due to irrigation is:

$$\Delta profit = profit_{irr} - profit_{noirr} \quad \begin{cases} > 0 & \text{irrigation is profitable} \\ \leq 0 & \text{irrigation is not profitable} \end{cases}$$

where, *c.f.* is a conversion factor, which is required to convert crop yields in ton/ha predicted by SWAT into yields in bushel/ha. The crop prices, production costs and conversion factors we used in profitability computation are listed in Table 4.5.

The Mackinaw River Basin possesses diverse soil types. Soil type distributions are shown in Figure 4.3. The soil data with BASINS 3.0 software package are from the State Soil Geographic (STATSGO) database developed by the National Cooperative Soil Survey. The data are suitable for the analysis at a regional scale. Obviously, the irrigation trigger values used in the simulation will affect the results. In our study, through a large number of experimental runs, profit-maximizing optimal trigger values for each soil type over thirty years are determined and used in the subsequent production runs.

Note that in the trigger optimization process we exclude the irrigation capital costs from our profits calculation.

In profits calculation, constant price and cost figures derived from historical records are used. However, in reality the higher crop yields require more fertilizer to be applied. In our simulation, we assume farmers are always able to apply sufficient fertilizer to meet the demands of crop growth. But our profits calculation neglects the related cost change. Moreover, in the long run, climate change may affect the relationship between supply and demand in markets and hence crop prices. Development of technology will also affect the costs. To project such changes in crop prices and production costs far exceeds the scope of our study. Hence, in this sense the economic analysis is more qualitative than quantitative.

Table 4.5   Crop market prices, production costs and unit conversion factors

Crops	Corn	Soybean	Winter Wheat
Price ($/bu)	2.47[1]	6.02[1]	3.10[1]
Production costs ($/ha-yr)	586.6[2]	387.4[2] (single crop) 213.07[3] (double-crop)	324.5[2]

Notes:
1. The average prices received by farmers in Illinois during the period 1991–2000, computed from the data in 2001 Illinois Annual Summary, Illinois Agriculture Statistics Service.
2. Computed from the survey data of the USDA costs and returns estimation program, the opportunity cost of unpaid labor and opportunity costs of land are excluded.
3. Computed from Estimated Costs of Crop Production in Illinois, 2003, Farm Business Management Handbook FBM-0100, Department of Agricultural and Consumer Economics, University of Illinois, 2003, opportunity costs of land are excluded.

Simulation Results

This section presents the simulation results and relevant analyses of agricultural productivity, profits and the low flow frequencies at references gauging stations under various formulated scenarios. Given that the double-cropping of winter wheat and soybean is not successful under the current (Illinois) climate in Mackinaw River Basin, only the results about double-cropping in the Oklahoma climate pattern are presented.

The impacts of climate change on productivity
Figure 4.4 shows the direct effects of climate change on the yields of three alternative crops. Annual average crop yields over the 30-year simulation period are computed and plotted in no irrigation cases and for two different CO_2 concentration levels in the Oklahoma climate pattern.

When the CO_2 concentration is held constant at 350 ppm after climate change, the yields of corn and soybean decrease by 20 percent and 18 percent respectively from yields under the current climate. However, simulation results indicate such adverse effects of climate change will be overcome by the increased CO_2. Simulation results also suggest that there will be a 16 percent increase in corn yield and a 31 percent increase in soybean yield if CO_2 is doubled to 700 ppm under the Oklahoma climate pattern.

Similar observations can be made about the yields of winter wheat and double-crop soybean. Under the Oklahoma climate conditions, their yields will increase by 58 percent and 53 percent respectively if CO_2 is doubled.

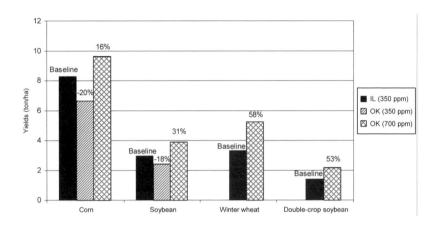

Figure 4.4   The impacts of climate change on productivity

The impacts of climate change on profits

The impacts of climate change on the agricultural production profitability and the economic viability of irrigation are demonstrated in Figure 4.5. Profits of producing crops in every year of the simulation period are evaluated in terms of dollars per hectare and plotted in the form of a histogram. The lines in the histograms represent the empirical cumulative distributions of profits of producing these crops. In the profits computation for Figure 4.5, irrigation capital costs are excluded. Because the optimized irrigation trigger values are used, there are always some increases in profits as they are shown.

Among the three alternative crops, profits of corn planting exhibit larger increases to irrigation. In the Oklahoma climate pattern and at $CO_2 = 350$ ppm, the annual average profit increase of corn production over the entire watershed is as high as $143/ha (not including irrigation capital costs). Given the diversity in soil properties and the difference in costs between surface water and groundwater irrigation, analysis indicates that irrigation will probably occur in a portion of the watershed, especially in riparian land with surface water irrigation. However, at $CO_2 = 700$ ppm, the effects of elevated CO_2 reduce the water demand of crops and make the profits from irrigation decrease significantly. In the case of winter wheat, the two lines representing the profit distribution in no irrigation and irrigation cases are almost overlapping.

Given the fact that most areas of the Mackinaw are not irrigated, it could be more appropriate to use the profits, including the irrigation capital costs, to judge the economic viability of the irrigation. The agricultural profitability under various scenarios, which include the irrigation capital costs in profits computation, are not shown in Figure 4.5. After irrigation capital costs are included, the lines that represent the profits in irrigation cases shift left.

The impacts of climate change on regional water resources
The low flow frequencies at the Congerville gauging station under various scenarios are shown in Figure 4.6, also in the form of a histogram. Low flow frequencies, which resulted from alternative dominant crops, are plotted in different panels. In each panel, one of the lines plotted represents the low frequencies under current Illinois climate (labeled as 'IL') and others represent the low flow frequencies under various conditions with the Oklahoma (OK) climate pattern. Those lines that are overlapping are plotted as a single line and labeled together.

Irrigation will influence stream flow either by producing runoff or changing the antecedent soil moisture condition. Groundwater irrigation and surface water irrigation may have opposing effects. While abstraction of water from streams in surface water irrigation tends to decrease stream flow, groundwater irrigation from deep aquifers often augments stream flow (Linsley et al. 1982). Due to the augmenting effects of groundwater irrigation, it is possible that the irrigation actually decreases the low flow frequency.

In Figure 4.6, the lines labeled as 'irr–GW+SW' are produced to represent the cases in which both groundwater and surface water irrigation occurs in the entire watershed. On the other hand, the lines labeled as 'irr–SW' represent the low flow frequency in the case in which only surface water is applied in the riparian area. Intuitively, these two lines can be roughly considered as the two limits of streamflow frequency that could occur in the case of irrigation. If irrigation only occurs in some area of the watershed, for example, due to an assumption of the unlimited availability of groundwater from the deep aquifer which does not hold in reality, it is expected that low frequency representing this case will fall within the two bounds.

Similar to what is observed in the histogram of profits, the elevated CO_2 has significant effects on watershed hydrological conditions. In the OK climate pattern after climate change, if CO_2 is still at 350 ppm, the number of low flow days at the two reference stations will increase. However, if CO_2 increases to 700 ppm, the increases of low flow frequency will not be so significant, if any. In fact, for single soybean and winter wheat and soybean double-cropping, the low flow days almost vanished given the current $_7Q_{10}$ low flow standards.

Various simulation results on low flow frequency in OK climate patterns are presented. However the likelihoods of their occurrence are not equal. Although we can not exclude other possibilities with 100 percent certainty, our analysis of the economic impact enables us to choose the low flow frequencies occurring at 700 ppm in no irrigation cases as the scenario with the greatest likelihood.

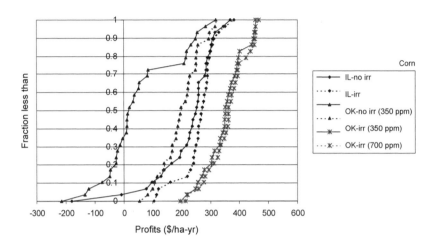

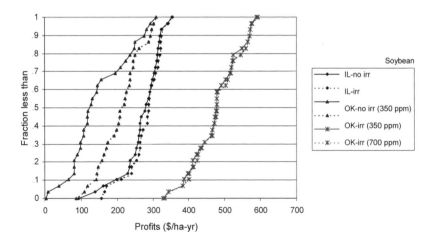

(figure continued on next page)

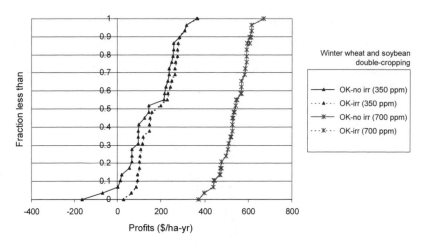

Figure 4.5   The impacts of climate change on agricultural production profitability (not including irrigation capital costs)

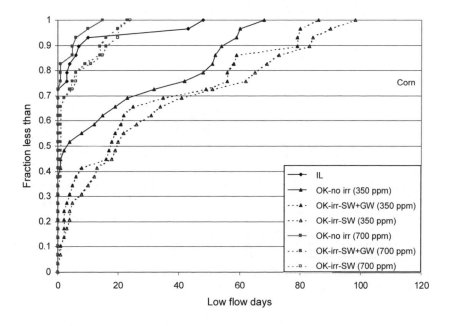

(figure continued on next page)

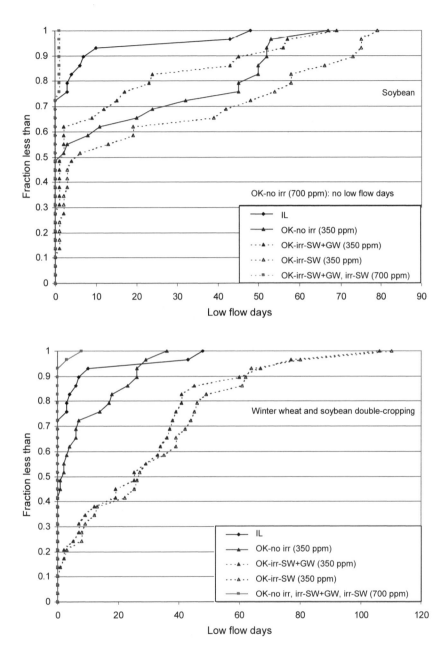

*Figure 4.6   Histograms of low flow frequency at the gauging station near
            Congerville*

Summary of Findings

The simulation and analysis results indicate that the effect [single] of elevated CO_2 is a key factor to determine the appearance of the watershed in the future. If CO_2 is held constant at the current ambient level, 350 ppm, changes in temperature and precipitation tend to decrease the crop yields and increase the low flow frequencies at the two reference gauging stations near Congerville and Green Valley. Among the three alternative crops included in this study, corn exhibits larger response of profits increase to irrigation. Economic analysis on irrigation effects in corn production indicate irrigation probably will be needed at some areas of, if not the entire, watershed. However adverse effects arising from the changes in temperature and precipitation will be overcome by the effects of elevated CO_2. Although there exist various possibilities, the following scenario is the one with the most likelihood. At doubled $CO_2 = 700$ ppm, the agricultural productivity will be improved. For example, the yields of corn and single crop soybean will increase by 16 percent and 31 percent compared to their yields under current Illinois climate. The elevated CO_2 also decreases the water demands in crop growth and hence reduces the potential irrigation needs. In that case, the watershed can maintain the traditional rain-fed agriculture. There will be fewer low flow days given current low flow standard $_7Q_{10}$ at the two reference stations. The vulnerability of the regional watershed will be ameliorated.

ECONOMIC IMPACT ANALYSIS

The IPCC (1998) has noted that climate change will affect regional economies differently according to their sectoral composition. It has been speculated that agriculture in the Midwest could be particularly susceptible to climate change. In investigating possible effects, Dixon and Segerson (1999) have projected that moderate temperature increases are likely to cause net revenue losses, although these losses will begin to decrease as the change in temperature increases from 2.5°C to 5°C. An increase in extreme meteorological events may force farmers to adopt alternative agricultural practices. In another study (cited above), the National Assessment Synthesis Team of the US Global Change Research Program predicts that corn yields are likely to decrease in the southern portions of Indiana and Illinois by 10–20 percent (NAST 2000b). The historical relationship between Midwest soybean yields and precipitation suggests that yields will decrease if either extremely wet or extremely dry conditions become more prevalent. The 1988 drought resulted in a crop yield decrease of some 30 percent, as did the flood of 1993. Dixon and Segerson (1999) predict the maximal dollar losses to Midwest agriculture from increased rainfall for a given temperature will range between $1.4 billion and $2.1 billion per year.

In this section, we discuss research on possible short- and long-term impacts that climate change can have on the economy of the six-county region of the Mackinaw River Basin in Illinois with particular emphasis on agriculture, transportation and utilities. We conduct impact analyses with an econometric input–output model of the region making use of simulation results obtained with the river basin model discussed in the previous section. Manifestations of climate change we consider include extreme weather events, the increased incidence thereof, long-term increases in greenhouse gas emissions and average seasonal temperatures, and changes in precipitation patterns.

Impacts of extreme, or what Barclay Jones termed 'unplanned', events have been studied extensively with input–output and CGE models (see Jones 1997; Okuyama 1999; Rose et al. 1997) and more recently, with linked transportation and input–output models (Cho et al. 2001). Such models are for the most part static frameworks that are best suited to considering the effects of once-and-for-all changes via a sequence of 'snap shots' of an economy. Nordhaus and Yang (1996), among others, have examined potential effects of long-term systemic climate changes at the multi-state regional level but at a high level of sectoral aggregation. To capture the evolving nature of climate change, involving both an increased incidence of extreme weather events and long-term systemic changes, and to assess its possible impacts on regional economies with a degree of sectoral disaggregation that permits insight into structural change, a more suitable approach to take might be to conduct simulations with regional econometric input–output models (or REIMs). Such simulations might be driven by solutions of ecological models based on alternative assumptions of changing climate conditions (see Conway 1990; Israilevich and Mahidhara 1991; Brodjonegoro 1997 for relevant examples of REIMs). Rey 1998 provides a comparative assessment of different types of REIMs, and examples of suitable ecological models would include the US EPA's BASINS suite of models and USDA's SWAT model. This is the approach we take in the present study.

While regional economic impacts of climate change at the national and multiple-state level of aggregation have received considerable attention (for example Nordhaus and Yang 1996; IPCC 1998; Yohe and Schlesinger 2002), impacts at the sub-state level have not. The study by Rubin et al. (1995), who carried out a comparison of impacts of climate change on the rural recreation-based economy of the Pere Marquette watershed and the industry- and commercial-activity-based economy of Milwaukee, is therefore notable. Rubin et al. (1995) found that climate-induced economic impacts on their study area exist and can be identified through an integrated econometric economic-environmental-energy demand modeling framework. The variables most affected in simulations conducted with models calibrated for Pere Marquette and Milwaukee were farming, service and retail employment levels, per capita income, net migration and park attendance.

The Mackinaw River Basin Economy

The contiguous Illinois counties of Ford, Livingston, McLean, Mason, Tazewell and Woodford, which contain the Mackinaw River Basin, cover an area of 4430 square miles. The actual watershed area occupies approximately 25 percent of the six-county area. In the year 2000 the population of the six-county region was 375 000 and the average household income was approximately $65 000. Agriculture is a relatively important sector in the region's economy, accounting for 2.4 percent of gross regional product (GRP) and 5 percent of total regional employment in 1998. Agriculture's shares of GRP and regional employment were twice that of this sector's shares of gross state product (GSP) and total state employment. In 1998 manufacturing's share of GRP was about 20 percent (slightly higher than the sector's 17 percent share of GSP). Finance, insurance and real estate's share of GRP was 24 percent (compared with 20 percent of GSP) and the service sector's share of GRP was 17 percent (compared with 22 percent of GSP). Government accounted for 10 percent of GRP and 12 percent of regional employment, whereas wholesale and retail trade accounted for 15 percent of GRP and 20 percent of employment. Other sectors' 1998 shares of output or employment were not significantly large and were approximately the same for both the six-county region and the state of Illinois.

In spite of the prominence of agricultural land use, the regional economy has a diversified economic base with firms doing business in 245 (of 528 classified) 3-digit SIC industries. The size of the workforce in many industries is small however there are only 148 industries with 50 or more employees and 128 industries with 100 or more employees.

Construction, Calibration/Estimation and Validation of the REIM

The basic structure of the regional econometric input–output model we employ closely follows that of Israilevich and Mahidhara's (1991) model of Chicago and Brodjonegoro's (1997) model of Jakarta, with one major exception – intraindustry wages and changes in demographic characteristics are not determined endogenously. In addition to general equilibrium conditions, interaction among sectors, and an intertemporal (or time-series) orientation, REIMs usually feature highly disaggregated sectoral levels. Because of the small number of sectors with large workforces in the study area, we consider it appropriate to aggregate up to only nine sectors:

agriculture; mining; construction; transportation and utilities; manufacturing; finance, insurance and real estate; services; trade; and government.

The core of the model consists of nine industry equation groups. Each group characterizes production relationships within an industry and how each industry's production process interacts with other sectors of the economy. Each industry grouping contains four equations determining two output levels (predicted and actual), employment and income (earnings or total factor payments). The first of the four equations determines predicted output according to the standard accounting identity for total sales from the corresponding industry row of an input–output table for the six-county region,

$$Z = \sum_{j=1}^{9} a_{jj} X_j + \sum_{k=1}^{4} b_{ik} FD_k \qquad (4.3.1)$$

In (4.3.1), Z_i is the output predicted for industry i, X_j is the actual (or observed) output for industry j, FD_k is the kth component of final demand, a_{ij} is the value of sales from industry i to industry j per unit of industry j's output, and b_{ik} is the value of sales from industry i to the kth component or source of final demand per unit of demand. Sources of final demand for industrial output included in the model are household consumption, investment, exports and non-local government. Equation (4.3.1) is deterministic and, apart from the benchmark year for which the regional input–output table has been constructed, predicted output will not equal actual output.

The second equation of the industry block is an 'output-correction equation', which accounts for the relationship (discrepancy) between predicted and actual output in any given year. It also characterizes the nature of technological change within the industry. The form of the output-correction equation in the model,

$$X_i = Z_i \exp f\big((\cdot) + \varepsilon_i\big), \qquad (4.3.2)$$

follows from

$$\log(X_i / Z_i) = f\big((\cdot) + \varepsilon_i\big), \qquad (4.3.2')$$

in which $f(.)$ is a function of exogenous and endogenous variables that account for technical change and ε_i is a stochastic error term. The relationship depicted by (4.3.2) has been found to be well-supported empirically, because the prediction error of an input–output table for out-of-sample (or non-benchmark) years is generally systematic – not random (see Brodjonegoro 1997). The third equation of the industry block,

$$N_i = X_i / (\exp(g(\cdot) + \mu_i),$$ (4.3.3)

determines the size of the workforce in industry i, N_i, and is based on a relationship in which changes in productivity (output per worker), X_i/N_i, are systematically accounted for by a function of exogenous and endogenous variables, $g(.)$ and random factors, μ_i,

$$\log(X_i / N_i) = g(\cdot) + \mu_i$$ (4.3.3')

The fourth and final equation of the industry block determines total factor payments or income in an industry. It describes the relationship between an industry's income, Y_i, and output, X_i, in terms of changes in exogenous and endogenous variables, included in $h(.)$ and random influences, ξ_i,

$$Y_i = X_i \exp(h(\cdot) + \xi_i)$$ (4.3.4)

The specifications for the three econometric (stochastic or non-deterministic) equations are not homogeneous across industrial groupings, but explanatory variables in output equation (4.3.2) generally include sectoral employment in the region, sectoral employment in the state, and the state level of sectoral product. Explanatory variables in employment equation (4.3.3) generally include lagged values of sectoral employment and output, the state level of sectoral employment, and the regional population. Explanatory variables in income equation (4.3.4) generally include lagged values of industry employment and income, industry income in the state and the regional population. Some equations, such as those in agriculture, include other exogenous variables, such as average wage, government payments to all industries, government payments to farms, farm labor income and a time trend. (for complete details see Donaghy and Plotnikova 2003).

The model is closed with three adding-up requirements for output, X, employment, N, and income, Y, in the regional economy,

$$X = \sum_{i=1}^{9} X_i,$$ (4.3.5)

$$N = \sum_{i=1}^{9} N_i,$$ (4.3.6)

$$Y = \sum_{i=1}^{9} Y_{i.} \qquad (4.3.7)$$

The interindustry sales (regional purchase) coefficients in the accounting identity (4.3.1) were obtained from the 1997 transactions matrix for the six-county region generated by the IMPLAN Pro (version 2.0) program and IMPLAN's Illinois county database. The final demand coefficients were obtained from the IMPLAN 1997 final demand output for each sector by dividing the demand for each sector by totals demanded of households, government, capital investment and net exports.

Time series data on final demand were constructed for each sector using Illinois data from the Illinois REIM model developed by the Regional Economic Applications Laboratory (REAL) at UIUC and the Federal Reserve Bank of Chicago. Appealing to the assumption of constant-returns-to-scale technology implicit in the model, data on industry output, employment, income and final demand for the six-county area were derived by multiplying the state values for a sector by the ratio of the six-county area employment to Illinois employment for that sector. Data on sectoral employment, income, government payments, average wage and population for the six counties were obtained from the online Regional Economic Information System (REIS) database. The availability of time-series data on sectoral output limited the sample to 22 annual observations corresponding to 1977–1998.

The model was estimated in three blocks of nine equations (output, employment and income) with the econometric software package Time Series Processor (TSP) version 4.4. A nonlinear three-stage least squares (NL3SLS) estimation approach was chosen over autoregressive (AR1 and AR2) linear estimation, commonly employed in the estimation of other REIM models, because the former approach enabled theoretical restrictions to be imposed across equations. The homoskedastic-consistent standard-error option for estimation was employed. While it is difficult to test directly for the presence of heteroskedasticity and autocorrelation in the error terms of a nonlinear simultaneous equation model with lagged endogenous variables when using the NL3SLS estimator in this package, the presence of these conditions would not affect the degree to which the parameter estimates themselves are unbiased or consistent, nor would it impair the use of the model for forecasting and simulation work, which is our primary concern. (That is, we are not testing hypotheses about the values of the parameters.) The fit of most equations of the estimated model was very good, with single-equation R-square values above 0.90 and proportionate root-mean-square errors less than 0.10 in 21 of 27 equations.

To explore the plausibility of the model's out-of-sample performance, forecasts were made for 25 years beyond 1998, the last year for which data were available. Model solutions were based on assumptions that the model's exogenous variables will evolve in a pattern that is in keeping with past historical trends and cycles. Plots of output and employment variables for key sectors suggest that the model's solution is well-behaved for historical variation in exogenous variables and plausible shocks to endogenous variables. Taken together with the model's in-sample performance, the out-of-sample forecasts suggest that the model will support plausible inferences drawn from simulations conducted with the model.

Impact Analyses (Simulations)

The purpose of the simulations here discussed is to provide approximations of effects of climate-induced changes on individual sectors and on the economy as a whole. Each simulation is intended to reflect a different climate change scenario. The simulations conducted include the effect on the regional economy of extreme events in the form of icestorms, through their impact on the transportation and utilities sector and long-term systemic increases in temperature and CO_2 and changes in patterns of precipitation, through their impacts on agriculture.

Extreme events such as heavy snowfalls and icestorms disrupt transportation networks and power distribution grids. According to Curtis Hawk, Assistant Director at the Emergency Disaster Services in Bloomington, Illinois, losses of up to 33 percent of capacity in major utilities lasting for two weeks occur, on average, every ten years. Taking the baseline solution of the model from 1978 to 2023 to be the first simulation, we consider in the second simulation what the effects of 33 percent losses in utility capacity of one month in duration might have been if they occurred in 1978, 1987 and 1997, and then onwards to 2023.

From the plot of simulation output data against the historical data (Figure 4.7), one can see that for the in-sample period, the transportation and utilities and trade sectors would have been significantly affected. Although not shown here, the capacity losses would most likely have contributed to modest short-run reductions in employment, income and production in the regional economy. This pattern continues out of sample. For this scenario we also explored the effect of remedial action by the government in response to transportation/utilities capacity loss by increasing government sector demand by 3.5 percent in the year following the year of transportation/utilities capacity loss. The offsetting effect of such action would likely more than offset the negative impacts of the capacity losses.

*Figure 4.7   Out-of-sample solution for transportation output, with and
without capacity loss*

Having considered sectoral and economy-wide impacts of episodic events associated with climate change, we turn now to simulations that take into account long-term systemic changes. According to the IPCC (1998), the direct effects on crop production of increases in temperature and CO_2 are likely to be largely beneficial, as CO_2 in the atmosphere acts as a natural fertilizer. The IPCC expects a doubling of CO_2 to increase crop yields of cotton, soybean and wheat by an average of 30 percent with a range of variation of -10 percent to +80 percent. Using the 1930s local (Bloomington measuring station) climate data, which are similar to conditions forecasted for the next few decades, Wayland Eheart and colleagues in the Department of Environmental Engineering at UIUC have employed the USDA's SWAT model to simulate how different weather patterns and an increase of CO_2 (from 350 ppm to maximum 700 ppm) over the 30-year period of 1963–93 would have affected corn yields and farm profits in the study region. In the third simulation, we incorporated the induced changes in crop yields implied in the work of Eheart in the product data and resolved the model. The changes were incorporated into the model by multiplying the agricultural output by the ratio of corn yield under the scenario simulated and corn yield under base scenario, weighted by the share of corn in agricultural output of the six-county area. We considered four scenarios:

1.   a four-year spell of excessive precipitation;
2.   a four-year spell of dry weather;
3.   a gradual increase in CO_2 in the atmosphere; and
4.   a gradual increase in CO_2 coupled with excessive precipitation.

The period of time considered in the first two scenarios is 1978 to 1999, whereas in the last two scenarios the time period is 1978 to 2022. The graphs in Figures 4.8–4.11, which depict the difference between the model solution with and without the manifestation of climate change in the respective scenario, indicate that an increase in agricultural output would have been or would be likely in all cases but that of the four-year drought. Employment and income in the agricultural sector and in the aggregate regional economy (not shown) would have been similarly affected.

Conclusions

In this summary we have reported on some preliminary investigations of possible short- and long-term impacts that different manifestations of climate change might have on the six-county regional economy of the Mackinaw River Basin in Illinois. The work to date has focused on the construction and validation of a regional econometric input–output model and its implementation in a limited number of simulations, some of which employ solutions of an ecological model. The simulations conducted have been exercises in counterfactual analysis, in which we have 'replayed' climatological and economic history, and 'what-if' analyses of future scenarios.

The simulations suggest that individual extreme events of the kind associated with climate change, or a systematic pattern of such events, will have significant short-term impacts at the regional level that may last several years beyond their occurrence. They would not appear to pose a threat of serious structural damage to the regional economy, however. Increases in temperature and CO_2 would appear to favor the agricultural sector and complementary sectors. (Owing to certain limitations of its specification, the model is not sensitive to how increases in temperature and CO_2 would affect other sectors. Hence, it is mute about the implications of the same. This shortcoming suggests the need for more comprehensive integrated assessment studies in which impacts on human health and ecosystems, *inter alia*, are also considered.)

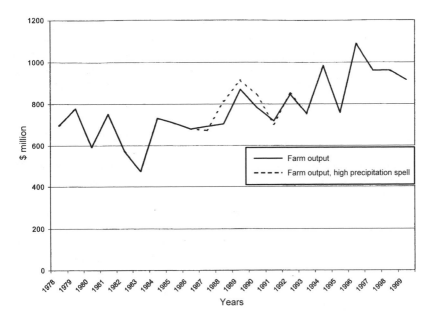

*Figure 4.8   Solution for agricultural output, with and without high
             precipitation*

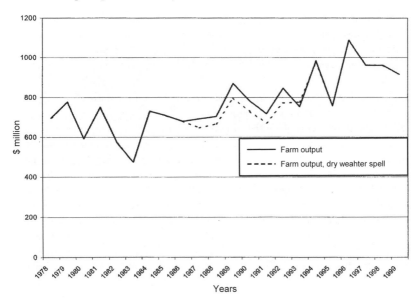

Figure 4.9   Solution for agricultural output, with and without dry weather

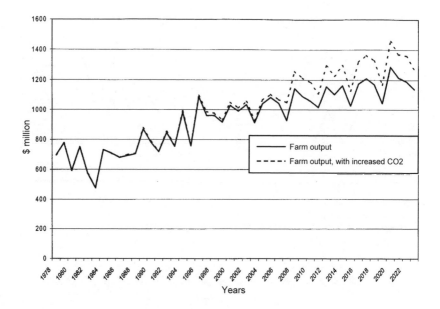

Figure 4.10   Solution for agricultural output with and without increased CO₂

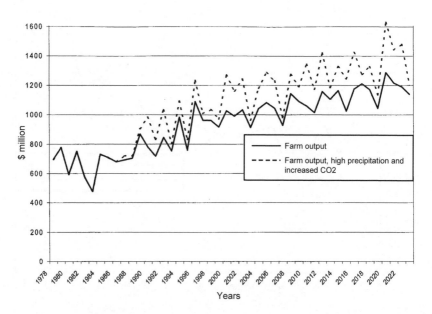

*Figure 4.11   Solution for agricultural output, with and without high
precipitation and increased CO₂*

NATURAL RESOURCES ASSESSMENT: FISH POPULATIONS

This research considered long-term consequences of climate change on natural resources, particularly fisheries. In this analysis, modeling tools were also used to predict the changes in fish populations with changing flow conditions. The approach used a habitat simulation model and a population model based on available habitat. Using this model system, it is possible to evaluate the long-term effects of flow change and use historical flow scenarios to predict the effects/impact of climate change on a critical ecosystem element.

The challenges for watershed management, when considering climate change issues, include choice of climate change scenario, evaluation of climate change effects, integration of climate change effects with other mechanisms of change, and finally development of management programs that operate effectively for the changes produced and over the time scales associated with climate change events. It is of concern that the assessment tools commonly used in watershed management may provide inadequate support for the management challenges associated with climate change. For example, the progressive environmental alteration associated with climate change will require process-based assessments that effectively track change in environmental conditions while illuminating mechanisms of alteration. Unfortunately, existing assessment tools fall short of meeting process-based criteria. For example, existing biomonitoring approaches provide an excellent means of assessing an existing condition or state, and compilations of past biomonitoring data do support trend analysis and provide limited support for prediction. In summary, traditional biomonitoring approaches are largely descriptive, applicable only to the time of collection, are often affected by an uncertain history, and often fail to define cause and effect when complex environmental settings are encountered (Herricks 2001). What is needed is the development of biomonitoring tools that are process-based, rather than descriptive, that support analysis over time scales appropriate to climate change, considering both history and future performance, and tools that more adequately define cause and effect in complex and changing multiple stressor environments.

This summary reviews the development of a prototype tool, which is process-based, and provides a process-based assessment using accepted population modeling approaches. The tool described integrates two stressors and has the capacity to address time-related change using both historical and predicted environmental conditions. We do not argue that the analysis/assessment tool developed necessarily meets all of the challenges of watershed management, but we feel the approach does provide a process-based assessment approach that considers long-term trends, identifies mechanisms of change, and provides insight into biological processes that better define cause and effect in complex, long-term, multiple stressor, assessment problems.

To assess possible impacts of any climate change scenario, it is necessary to develop a method of quantifying change and evaluating those changes in terms of impact criteria. Impact criteria are developed from performance measures. For example, performance measures for watershed management might include regulatory effectiveness, economic efficiency, equity, administrative ease and robustness/flexibility. Similar performance measures for natural resources, specifically ecosystem protection, would include population characteristics, community stability, or ecosystem processes. The relationship between the performance measures and the actual desired characteristics is the basis for an impact assessment. We have found that developing performance measures of ecological systems benefits from the adoption of a uniform set of definitions. The following definitions, developed by Eheart and Tornil (1999), are generally consistent with accepted definitions for water resources (Hashimoto et al. 1982a, b; Fiering 1982a, b, c, d) and ecosystems (Westman 1978): vulnerability reflects the severity of failure; resilience reflects the ability of the system to 'bounce back' from a failure event; reliability reflects the probability of meeting a standard; robustness reflects the insensitivity or satisfactoriness of the performance variable in the face of parameter uncertainty; and recovery reflects the time between failure events.

By using vulnerability, resilience, reliability, robustness and recovery measures, it is possible to define impact criteria that are appropriate to climate change analysis. For example, vulnerability should be assessed both in terms of absolute change in an assessment parameter, and the time period over which that failure event extends. Resilience is similarly assessed in terms of an absolute change in the assessment parameter, and the speed (time related change) with which that parameter approaches pre-failure conditions. Reliability accepts the fact that there will be variation in assessment parameters and that it is essential to have both a measurable standard and a reasonable expectation of what the standard should be. Robustness can be assessed in the lack of variation in an assessment parameter. Recovery can be determined both in an absolute sense by comparing pre- and post-event measurements, and the time-related issues of parameter change occurring between failure events.

Assessment Tool Development

The assessment tool developed to provide a basis for prediction of multiple stressor (hydrology/flow and water temperature) effects of climate change required the integration of several models. The core model is the Physical HABitat SIMulation (PHABSIM) procedures developed by the US Fish and Wildlife Service and presently maintained by the US Geological Survey.[3]

PHABISM uses discharge data with representative reach cross section depth, velocity and substrate data to provide stage specific estimates of channel hydraulic conditions. These hydraulic data are integrated with species and life stage specific habitat suitability information to produce an estimate of habitat quantity. This habitat quantity metric, the weighted usable area (WUA) is developed for a representative reach, which can be extrapolated to other reaches in the drainage net with similar channel characteristics. PHABSIM was applied to a reach of Panther Creek, a major tributary of the Mackinaw River in Central Illinois. Flow data were developed from USGS gauging stations providing a 50 year flow record and habitat assessment period (Bergner 2002). This approach allows analysis of historical flow conditions for model testing and the evaluation of modified flow regimes typical of climate change scenarios. Flows for climate change scenarios were developed using two methods. By the first method, historical records are evaluated and a flow sequence typical of years that were hotter or drier is constructed. The second method requires output from SWAT modeling (Wollmuth and Eheart 2000), which is driven by rainfall inputs of expected climate change scenarios.

The WUA estimates from PHABSIM were then used in a fish population model. Food and space are considered as major factors/limits to population growth. In this tool, space is estimated from WUA values for each life stage, and changing space/area availability associated with flow is used in the prediction of populations over time. Food is assumed to be non-limiting in the model. The population model was implemented in STELLA, a dynamic modeling environment. For this prototype tool a smallmouth bass (*Micropterus dolomieui*) model was used (see Figure 4.12). As illustrated in Figure 4.12, an optimum, weekly minimum and average WUA was determined. The optimum WUA was set at the highest WUA in the record; weekly minimum and average WUAs were determined from PHABSIM model output. The model was developed using life history data available from the literature, and species response data from Illinois, producing estimates of life stage specific population values.

A second input to the population model is water temperature. A nonlinear logistic stream-air temperature regression model was used to estimate water temperature from air temperature (Mohseni et al. 1999). The air temperature data were developed from historical weather data compiled and analyzed by Kunkel (personal communication 2000). This historical analysis provided both actual historical temperature conditions and an analysis that determined which years had temperature conditions that matched climate change scenarios (for example 1, 2, 5, 8°F increase over average conditions). In the model application, daily temperatures from years that provided scenario conditions were used to assemble a long-term data set to drive the water temperature model. The temperature estimates were used as a threshold value, identifying the onset of spawning for smallmouth bass (in this case the first six-day period in which water temperature ranged between 57 and 72°F). Spawning then continued for two months. The temperature model thus affected onset of spawning, independent of the suitability of flow conditions for spawning success.

The integrated modeling system provides an assessment tool that addresses critical issues of a process-based climate change assessment. Using this tool, it is possible to use fish populations as an indication of the effect of changing climate conditions. In this example, the relatively long-lived smallmouth bass is used, but other models have been developed for species with different life histories (Tompkins 1998). In this prototype application, the life history of the smallmouth bass provides a basis for analysis of the cumulative effects of annual conditions. Further, it is possible to analyze life-stage specific population responses to identify factors associated with population change. Through identification of factors required for population maintenance, it is possible to identify causal factors for population change, leading to identification of cause and effect relationships, critical in management practice selection. Although only two stressors were used in this prototype, the population model allows the evaluation of multiple stressors, when species response data are available.

Prototype Model Results

From model results it is possible to identify periods of spawning success and failure and to track adult population trends in relation to life state specific responses to flow. Using this prototype model, it has been possible to identify which historical flow conditions lead to predicted population change, and identify life stage specific characteristics associated with those changes. It is also possible to develop new indices for population model analysis that provide insight into population responses. Practical sampling limitations would prevent determination of these indices from field data. But it is possible, for example, to model changing juvenile population estimates and juvenile WUA, as well as spawning times and a fry/juvenile ratio.

Climate Change Impact Assessment

The prototype modeling tool provides a new approach to the measurement of vulnerability, resilience, reliability, robustness and recovery. For example, it is possible to identify vulnerability in terms of absolute population values as well as predicted population change over time. Further, it is possible to develop vulnerability indices, such as the ratio of existing population size to a target value, and then plot a change in these indices. It is also possible to conduct these analyses with single or multiple stressors, and consider a wide range of future conditions in the assessment. A similar approach can be used to analyze resilience, reliability, robustness and recovery, develop new indices for these measures, and apply those indices in historical and/or predictive analysis.

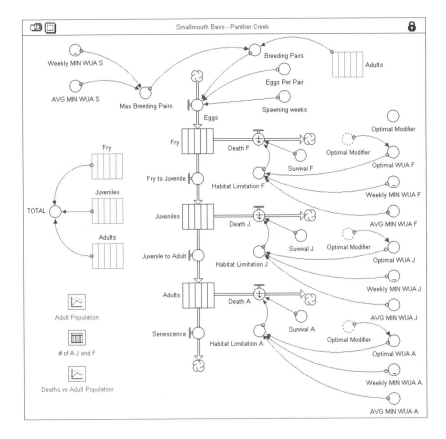

Figure 4.12   STELLA model of smallmouth bass population growth incorporating WUA and temperature limitations

Prototype Tool Assessment

An argument developed in this summary is that the future of water resources management will present challenges to scientists and natural historians that can only be met by developing process-based tools. Existing descriptive approaches used in biomonitoring and environmental analysis provide an essential foundation for assessment activity in watershed management, but the limits of descriptive approaches are becoming more apparent as management emphasizes performance and outcome.

The prototype tool identified in this chapter is limited in its present application, but it suggests promise in many areas. The major limit to this tool is effective calibration and sensitivity analysis. To accurately calibrate the model, long-term population information would be needed, information that is based on a response only to flow or temperature. The data are not available, although population model characteristics were evaluated against a long-term population record in the development of models for each species (Tompkins 1998). The tool does have an advantage in that the modeling structure is transparent in the sense that the integration of flow and temperature in the tool is achieved through a population model. Further, population model development has followed accepted population modeling theory that has been well-described (Pielou 1974; Nisbet and Gurney 1982). The implementation in STELLA provides a practical, and transparent, model structure as well. The end result is the addition of a single tool to the array of tools that can potentially be used in watershed management. With an array of tools available, the watershed manager is challenged in both selection and proper use of tools, independent of concerns about accuracy, precision or utility of any single tool.

Conclusions of Natural Resources Assessment

The prototype tool described in this summary provides a process-based approach to the assessment of long-term alterations in environmental conditions. The approach provides life stage specific population estimates, which supports the development of new indices of response, which can be tailored to impact characteristics such as vulnerability, resilience, reliability, robustness and recovery. The tool has application in a general DSS in that it can provide an example of effect/impact that can be generalized, even though the analysis is site specific.

EVOLUTION OF THE PROJECT DSS

Understanding and responding to regional impacts of climate change could be greatly facilitated by online (or worldwide web-based) decision (or planning) support systems (DSS) which can help communities in a common region engage the planning problem (for example Giaoutzi and Nijkamp 1993; Guhathakurta 1999). Generally speaking, DSSs bring together databases, models and software tools for querying and manipulating data through a graphical user interface to support solution of semi- or unstructured problems. Data may be accessed via hypertext, geographical information system (GIS), or dynamic models according to hierarchical, topical, spatial or temporal references. Data required for decision making may be qualitative as well as quantitative. Online DSSs offer certain advantages over systems that reside on, say, a single computer platform. They can be more easily updated and provide access to all resources available on the worldwide web. In particular, online DSS can:

1. provide links to international and regional databases on web pages (for example those of the EPA, NOAA, DNR or the IPCC) where climate change is discussed at various levels of detail from different scientific and political perspectives;
2. convey a concrete sense of the local nature of impacts of climate change by providing access to archived information on local histories of the experience of extreme events as recorded in newspaper accounts and meteorological data;
3. promote a collective understanding of what a community faces or how its residents might envision responding to impacts of climate change by providing access to digitized recordings of public or focus group meetings in which representatives of different stakeholder groups – such as public works managers, public utility managers, farm bureau and insurance industry representatives – participate;
4. support 'what-if' scenario analyses that will permit communities to consider alternative response strategies and their attendant costs and benefits;
5. facilitate asynchronous exchanges of opinions about changes that communities are willing to undertake or real-time discussions at public meetings.

Difficulties of Developing DSSs to Support Understanding and Response to Climate Change

For a DSS truly to support understanding and to respond to regional impacts of climate change several technological and social difficulties must be overcome. Models of different social and natural systems, based on different scales of temporal and spatial resolution, must be integrated. Depending on the computational resources they require, partial-equilibrium solutions may need to be made available in look-up tables for access by other models. Model outputs must be interpreted and validated. For all of this to happen, unprecedented dialogue and collaboration between different elements of the scientific community must be promoted. Data must be accessed from remote sites, exchanged between models, visualized and/or summarized for public consumption. Interfaces must be developed that seamlessly integrate models, databases, local and remote web sites, and which permit users of different backgrounds and interests to engage the issues of regional impacts of climate change at a comfort level that enables them to feel like a meaningful participant in deliberation about planning responses.

In this project we were able to take an innovative approach to providing stakeholders and investigators with access to raw and summarized information and especially historical climate records. We provided user-friendly access to much valuable background information about climate change and the project as well as access to other web sites where more detailed information could be obtained. We did not succeed, however, in developing a functioning DSS that adequately represented all project thrusts. Whereas we had intended to create an interactive system that could be accessed from a number of entry points and could provide comprehensive coverage of a range of questions, technical limitations prevented full-scale implementation. Consequently, we adopted a 'look-up' table approach that presupposed what questions users would pose. Unable to develop a capability to handle general inquiries in which users can construct their own scenarios for analysis, we found it necessary to pre-select scenarios. That said, we were successful in integrating some disparate elements of the research through the DSS, which is available on the Internet, and in the case of the econometric input–output model, providing the capability for users to conduct real-time impact analyses.[4]

EVALUATION OF THE OVERALL PROJECT

We discuss in the following paragraphs how we faired in meeting the objectives of the project. Drawing on available information we were able to define differences and similarities in impacts on the watershed and sectoral responses. We were able to identify causal relationships but also found that there is a high potential for the influence of indirect effects, which put in question some causal imputations.

We were partly successful in identifying intersectoral relationships, with success being limited by the compatibility of models employed. Through interaction of investigators, however, we were able to initiate analysis of cascades of effects and identify influential factors.

We did apply the impact analysis paradigm and achieved considerable breadth of coverage in the areas of agricultural management, economic, ecological, hydrological and water quality. We succeeded in identifying factors that would be altered by climate change conditions and effectively developed a host of predictive tools to assess both effect and impact. Some of our research findings were surprising, with the most significant example being that adding changes in CO_2 to the evaluation of agriculture/water resource relationships leads to conclusions different from earlier analyses that fail to consider such changes.

While we did not explore the full range of possible sector/element interactions and the potential for adaptation in these interactions, we did examine adaptive responses in the agriculture/water-resource analysis and the economic impact analysis. We also investigated possible adaptive responses with stakeholders in the workshops conducted early in the project.

The success in integrating project results with the DSS has been discussed in the previous section of the chapter. Overall, we believe that the research project discussed in this chapter has improved our understanding of regional impacts of climate change and contributed to the advancement of our capabilities to protect the environment and public health. The research has demonstrated the application of tools already developed for environmental management and newly developed tools for impact analysis that allow us to assess the potential effects of climate change on the Midwest in the case of a small river basin. The project has moreover made advances in understanding how water use policies might be changed in agricultural areas and how interrelated elements of the economy might adapt to climate change. Finally, the project has demonstrated the importance of long-term analytical capabilities to the management of a natural resource such as river fisheries.

NOTES

1. Hopkins (2001) refers to the *interdependence, indivisibility* and *irreversibility* of decisions in the face of *imperfect* foresight, the four I's of planning.
2. More information on the Illinois Global Climate Change Project is available at: http://dnr.state.il.us/orep/INRIN/eq/iccp/iccp.htm.
3. The core PHABSIM procedures are available at: http://www.mesc.usgs.gov/products/pubs/15000/15000.asp.
4. The project DSS is available at: http://www.imlab.psu.edu/ricc.

REFERENCES

Arnold, J.G., R. Srinivasan, R.S. Muttiah and J.R. Williams (1998), 'Large area hydrologic modeling and assessment – part 1: model development', *Journal of the American Water Resources Association*, **34**(1): 73–89.

Bergner, E.R. (2002), *Prediction of how Climate Change Induced Stream Temperature Alteration affects Smallmouth Bass Populations in a Small Midwestern Stream*, MS Thesis, University of Illinois at Urbana-Champaign.

Bingner, R.L. (1996), *Runoff Simulated from Goodwin Creek Watershed Using SWAT*, Transactions of the ASAE, **39**(1): 85–90.

Bowman, J.A. and M.A. Collins (1987), *Impacts of Irrigation and Drought on Illinois Water Resources*, ISWS/RI-109/87, Illinois State Water Survey.

Bowman, J.A. and B.C. Kimpel (1991), *Irrigation Practice in Illinois, Research Report 118*, Illinois State Water Survey.

Brodjonegoro, B. (1997), *The Econometric Input–output Model of Jakarta, Indonesia, and its Applications for Economic Impact Analysis*, PhD Dissertation, University of Illinois at Urbana-Champaign.

Brown, R.A. and N.J. Rosenberg (1997), 'Sensitivity of crop yield and water use to change in a range of climatic factors and CO_2 concentrations: a simulation study applying EPIC to the central US', *Agricultural and Forest Meteorology*, **83**(3–4): 171–203.

Brown, R.A. and N.J. Rosenberg (1999), 'Climate change impacts on the potential productivity of corn and winter wheat in their primary United States growing regions', *Climatic Change*, **41**(1): 73–107.

Brumbelow, K. and A. Georgakakos (2001), 'An assessment of irrigation needs and crop yield for the United States under potential climate changes', *Journal of Geophysical Research-Atmospheres*, **106**(D21): 27383–405.

Carraro, C. and A. Goria (1999), *Integrating Climate Policies in the European Environment: A Policy Report*, FEEM working Paper No. 46–99, Venice.

Cho, S., P. Gordon, J.E. Moore, H.W. Richardson, M. Shinozuka and S. Chang (2001), 'Integrating transportation network and regional economic models to estimate the costs of a large urban earthquake', *Journal of Regional Science,* **41**: 39–65.

Conway, W. (1990), 'The Washington projection and simulation model: Ten years of experience with a regional interindustry econometric model', *International Regional Science Review*, **13**: 141–65.

Dixon, B.L. and K. Segerson (1999), 'Impacts of increased climate variability on the profitability of midwest agriculture', *Journal of Agriculture and Applied Economics*, **31**(3): 537–49.

Donaghy, K.P. and M. Plotnikova (2003), 'Analyzing regional economic impacts of climate change with econometric input–output and ecological models', *Annals of Regional Science*, under review.

Easterling, W.E., N.J Rosenburg, M.S. McKenney, C.A. Jones, P.T. Dyke and J.R. Williams (1992), 'Preparing the erosion productivity impact calculator (EPIC) model to simulate crop response to climate change and the direct effects of CO_2', *Agricultural and Forest Meteorology*, **59**: 17–34.

Eheart, J.W. and D.W. Tornil (1999), 'Low flow frequency exacerbation by irrigation withdrawals in the agricultural Midwest under various climate change scenario', *Water Resources Research*, **35**(7): 2237–46.

Fiering, M.B. (1982a), 'A screening model to quantify resilience', *Water Resources Research*, **18**(1): 27–32.

Fiering, M.B. (1982b), 'Alternative indices of resilience', *Water Resources Research*, **18**(1): 33–40.

Fiering, M.B. (1982c), 'Estimates of resilience indices by simulation', *Water Resources Research*, **18**(1): 41–50.

Fiering, M.B. (1982d), 'Estimating resilience by canonical analysis', *Water Resources Research*, **18**(1): 51–7.

Friedmann, J. (1987), *Planning in the Public Domain,* Princeton, NJ: Princeton University Press.

Giaoutzi, M. and P. Nijkamp (1993), *Decision Support Models for Sustainable Development,* Avebury: Avebury Press.

Guhathakurta, S. (1999), 'Urban modeling and contemporary planning theory: is there a common ground?', *Journal of Planning Education and Research*, **18**: 281–92.

Hashimoto, T., J.R. Stedinter and D.P. Loucks (1982a), 'Reliability, resiliency and vulnerability criteria for water resource system performance', *Water Resources Research*, **18**(1): 14–20.

Hashimoto, T., J.R. Stedinger and D.P. Loucks (1982b), 'Robustness of water resources systems', *Water Resources Research*, **18**(1): 21–6.

Herricks, E. (2001), *Observed Stream Responses to Changes in Runoff Quality*, manuscript.

Hopkins, L.D. (2001), *The Logic of Making Plans for Urban Development*, Covelho, CA: Island Press.

Illinois Department of Energy and Natural Resources (1994), *A Climate Change Action Plan for Illinois*, Springfield, IL: http://dnr.state.il.us/orep/INRIN/eq/iccp/toc.htm.

Intergovernmental Panel on Climate Change (IPCC) (1998), *The Regional Impacts of Climate Change An Assessment of Vulnerability*, Cambridge: Cambridge University Press.

Israilevich, P.R. and R. Mahidhara (1991), 'Hog butchers no longer: twenty years of employment change in Metropolitan Chicago', *Economic Perspectives,* Federal Reserve Bank of Chicago, **15**: 2–13.

Izaurralde, R.C., N.J. Rosenberg, R.A. Brown and A.M. Thomson (2003)',Integrated assessment of Hadley Center (HadCM2) climate-change impacts on agricultural productivity and irrigation water supply in the conterminous United States – part II, regional agricultural production in 2030 and 2095', *Agricultural and Forest Meteorology*, **117**(1–2): 97–122.

Jones, B.G. (ed.) (1997), *Economic Consequences of Earthquakes: Preparing for the Unexpected*, Buffalo: National Center for Earthquake Engineering Research.

Kunkel, K. (2000) Personal communication with E. Herricks.

Liang, X-Z. (2003) Personal communication with J.W. Eheart.

Linsley, R.K, Jr, M.A. Kohler and J.L.H. Paulhus (1982), *Hydrology for Engineers*, New York: McGraw-Hill.

Mohseni, O.T., R. Erickson, H.G. Stefan (1999), 'Sensitivity of stream temperatures in the United States to air temperatures projected under a global warming scenario', *Water Resources Research*, **35**(12): 3723–33.

Nafziger, E.D. (2003), Personal communication with J.W. Eheart.

National Assessment Synthesis Team (NAST) (2000a), *Climate Change Impacts on The United States – Overview Report*, Cambridge, UK: Cambridge University Press.

National Assessment Synthesis Team (NAST) (2000b), *Climate Change Impacts on the United States: The Potential Consequences of Climate Variability and Change.* Cambridge, UK: Cambridge University Press.

Neitsch S.L., J.G. Arnold, J.R. Kiniry and R. Williams (2001), *SWAT 2000 Theoretical Documentation*, Soil and Water Research laboratory, USDA–ARS, Temple, TX.

Nisbet, R.M. and W.S.C. Gurney (1982), *Modeling Fluctuating Populations*, New York: John Wiley and Sons.

Nordhaus, W.D. and Z. Yang (1996), 'A regional dynamic general-equilibrium model of alternative climate-change strategies', *American Economic Review*, **86**: 741–68.

Okuyama, Y. (1999), *Analysis of Structural Change: Input–output Approaches*, Ph.D. Dissertation, University of Illinois at Urbana-Champaign.

Pielou, E.C. (1974), *Population and Community Ecology Principles and Methods*, New York: Gordon and Breach Publishers.

Rey, S.J. (1998), 'The performance of alternative integration strategies for combining regional econometric and input–output models', *International Regional Science Review*, **21**: 1–36.

Rose, A., J. Benavides, S.E. Chang, P. Szczesniak and D. Lim (1997), 'The regional impact of an earthquake: direct and indirect effects of electricity lifeline disruptions', *Journal of Regional Science*, **37**: 437–58.

Rubin, B.M., J.W. Greco and M.D. Hilton (1995), *A Multidisciplinary Comparison of the Relative Rural and Urban Economic Impacts Associated with Global Climate Change*, Paper presented at the 25th Annual Meeting of the Urban Affairs Association, May, Portland, OR.

Santhi, C., J.G. Arnold, J.R.Williams, W.A. Dugas, R. Srinivasan and L.M. Hauck (2001) 'Validation of the SWAT model on a large river basin with point and non-point sources', *Journal of the American Water Resources Association*, **37**(5): 1169–88.

Sharpley, A.N. and J.R.Williams (eds) (1990), *EPIC-Erosion Productivity Impact Calculator*, 1. Model Documentation (Tech. Bull.1768), US Department of Agriculture, Agricultural Research Service.

Singh, Krishan P., G.S. Ramamurthy and Il Won Seo (1988), *7-Day 10-Year Low Flows of Streams in the Kankakee, Sangamon, Embarras, Little Wabash and Southern Regions: ISWS Report 441*, Illinois State Water Survey.

Southworth, J., J.C. Randolph, M. Habeck, O.C. Doering, R.A. Pfeifer, D.G. Rao and J.J. Johnston (2000)', Consequences of future climate change and changing climate variability on maize yields in the midwestern United States', *Agriculture Ecosystems and Environment*, **82**(1–3): 139–58.

Southworth, J., R.A. Pfeifer, M. Habeck, J.C. Randolph, O.C. Doering, J.J. Johnston and D.G. Rao (2002), 'Changes in soybean yields in the midwestern United States as a result of future changes in climate, climate variability and CO_2 fertilization', *Climatic Change*, **53**(4): 447–75.

Stockle, C.O., J.R. Williams, N.J. Rosenberg and C.A. Jones (1992), 'A method for estimating the direct and climatic effects of rising atmospheric carbon dioxide on growth and yield of crops: part 1 – modification of the EPIC model for climate change analysis', *Agricultural Systems*, **38**: 225–38.

Tompkins, M.R. (1998), *Analysis of the Fisheries in Maintained Streams Using the PHABSIM Model and a Habitat Driven Population Model*, MS Thesis, University of Illinois at Urbana-Champaign.

United States Agency for International Development (USAID) (1998), *Climate Change Initiative*, Washington, DC.

Uribelarrea, M., S.P. Long, S.P. Moose and F.E. Below (2003), 'Nitrogen use impacts the growth response of corn to atmospheric enrichment of CO_2', *ASA-CSSA-SSSA Annual Meeting*, Denver, Colorado.

Van Liew, M.W. and J. Garbrecht (2003), 'Hydrologic simulation of the Little Washita River experimental watershed using SWAT', *Journal of the American Water Resources Association*, **39**(2): 413–26.

Westman, W.E. (1978), 'Measuring the inertia and resilience of ecosystems', *Bioscience*, **28**: 705–710.

Wollmuth, J.J.C. and J.W. Eheart (2000), 'Surface water withdrawal allocation and trading systems for traditionally riparian areas', *Journal of the American Water Resources Association*, **36**(2): 293–303.

Wright, J. and F. Benson (1983), 'Cost Comparison of Alternative Irrigation Systems', *Agricultural Economics Fact Sheet No. 34, Table 3, A Cost Comparison of Alternative Irrigation Water Supply Systems*.

Yohe, G. and M. Schlesinger (2002), 'The economic geography of impacts of climate change', *Journal of Economic Geography*, **2**: 311–44.

5 Ecological and Economic Impacts of Climate Change in Agricultural Systems: An Integrated Assessment Approach

J. Antle, S. Capalbo and K. Paustian

INTRODUCTION

The Intergovernmental Panel on Climate Change's (IPCC) Third Assessment Report concludes that, 'greater emphasis on the development of methods for assessing vulnerability is required, especially at the national and sub-national scales where impacts of climate change are felt and responses are implemented. Methods designed to include adaptation and adaptive capacity explicitly in specific applications must be developed' (IPCC 2001, p. 22). In this chapter we present an approach to assess ecological and economic impacts of climate change on agriculture, taking into account two critical factors that affect agriculture's vulnerability: the degree of adaptation by farmers; and spatial heterogeneity in biophysical and economic conditions that affect the ability to adapt. Our approach is based on the use of site-specific data coupled with spatially explicit ecosystem and economic models to simulate farmers' land use and management decisions in response to climate change. With this statistically representative sample of data, the models can be simulated to represent the impacts of climate change on the population of economic decision units, and results can be used to quantify the impacts on the population and can be statistically aggregated for policy analysis.

We demonstrate this approach in an analysis of climate change impacts in the dryland, grain-producing region of Montana. The coupled ecosystem and economic models are used to simulate the effects of climate change and CO_2 fertilization on a key indicator of agricultural sustainability, the stock of carbon in agricultural soils. The analysis shows that adaptation has substantial impacts on both soil carbon stocks and on the vulnerability of the agricultural producers to climate change. The analysis also shows that the simulated effects of CO_2 on crop productivity play an important role in determining the ecological and economic changes in the agricultural system.

MODELING AGRICULTURAL IMPACTS OF CLIMATE CHANGE

Economic models of behavior show that the magnitude of changes in management needed to adapt to climate change are determined by the ways that climate affects the production system (Antle et al. 2004). Spatial heterogeneity is a key feature of agricultural production systems and should play an important role in analysis of adaptation. In this section we discuss the methods used in the literature to study impacts of climate change on agriculture and their capability to incorporate these two key features. We then provide an overview of the integrated assessment approach used in this study.

Previous Studies of Adaptation to Climate Change

Since the early 1990s, there have been various studies of agricultural impacts of climate change (see IPCC 2001, Section 5.3.4.5 for an overview). In their review of the literature on climate change impacts on US agriculture, Lewandrowski and Schimmelpfennig (1999) categorize the methods used as a crop response approach or a spatial analogue approach. In the crop response approach, crop growth models are used to simulate effects of climate change on crop yields, and these yield changes are then used to assess economic impacts. Studies have made different assumptions about adaptation, including changes in cultivars, planting and harvest dates, crop mix and use of irrigation (Kaiser 1999). Two features of the studies in the literature are notable relative to the preceding discussion of adaptation.

First, crop response studies have analyzed adaptation by simulating the effects of a specified set of management options that could mitigate the effects of climate change. These modeling exercises are implemented by introducing management adaptations into crop growth models and then evaluating the consequences under the assumption that they were adopted. A limitation of this approach is that the simulated management adaptations are not endogenously determined within the models. While this approach does provide a way to measure sensitivity to adaptation, one cannot assume that these adaptations are the ones that would actually be implemented by economically motivated farmers. A second point is that the crop response studies utilize data for a limited number of 'representative farms' or experimental sites, and thus do not account for spatial variability in the population of farms in a region of analysis.

In the spatial analogue approach (see for example Mendelsohn et al. 1994; Darwin et al. 1995; Segerson and Dixon 1998) variations in historical climate conditions over a large geographic area are included in a statistical model to infer how climate affects the value of agricultural production. One argument for using this method is that the statistical model's parameters embed the various adaptations that farmers make in order to be able to produce crops at locations with different climates. Mendelsohn et al. (1994) argue that the crop response approach does not account for the various adaptations that farmers make, and thus would over-state the impact of climate on agriculture. Studies based on the spatial analogue approach indicate that adaptation can mitigate the generally negative impacts of climate change on the value of agricultural production in the United States by up to 50 percent.

A number of limitations to the spatial analogue approach have been noted in the literature as well. First, the use of spatially aggregated data introduces aggregation bias. A well-know consequence of spatial aggregation is to bias downwards the effects of spatial variability (Hansen and Jones 2000; Antle and Stoorvogel 2001). Thus aggregation limits the extent to which we can quantify spatial differences in adaptation associated with spatial variability. A second problem with the spatial analog approach is that statistical models are estimated with cross sectional data, and then used to infer the effects of climate variation over time, a procedure that requires the unjustified assumption that climate variability across space has the same effects as climate variability over time (Schneider 1997). Another limitation is that statistical models based only on historical data cannot be used to simulate the effects of future increases in atmospheric CO_2 on agricultural productivity. Various studies have found the effects of CO_2 fertilization to play a key role in the analysis of agricultural impacts of climate change (IPCC 2001). Finally, conventional econometric production models based on profit or cost functions are defined for interior solutions to continuous input choice problems, and cannot be used to represent non-marginal changes in the choice of production activities, land use and management that are likely to be the result of climate change.

Integrated Assessment Approach to Adaptation in Agricultural Production Systems

In this section we outline an approach to modeling agricultural impacts of climate change that utilizes features of both the crop response approach and the spatial analog approach discussed above, but goes beyond those approaches to overcome some key limitations. Our goal is to provide an approach to assess impacts of climate change and CO_2 fertilization that incorporates endogenous adaptation to climate change and that takes into account the effects of spatial variability in biophysical and economic conditions. Our approach, like the crop response approach, uses biophysical process models to represent the impact of climate change and CO_2 fertilization on crop production. Like the spatial analog approach, our approach uses econometric models to represent and simulate farmers' endogenous response to climate change. However, unlike the spatial analog approach, the integrated assessment approach utilizes the process-based information embodied in a biophysical model to simulate the effects of climate and CO_2, by linking the biophysical model's output to an econometric process simulation model. The econometric process simulation approach uses site-specific data in order to obtain information about the spatial heterogeneity of the population and its adaptation to climate and CO_2 changes, and can simulate both discrete and continuous production decisions.

An overview of the integrated assessment approach is presented in Figure 5.1. The left-hand side of the figure represents the components of the analysis based on the Century model (see Parton et al. 1994; Metherell et al. 1993), and the right-hand side represents the components of the analysis based on the econometric process simulation model (Antle and Capalbo 2001; Antle et al. 2004). At the top of the figure, the incorporation of spatial heterogeneity is indicated by the use of site-specific biophysical data and economic data to parameterize and simulate the two models. A key feature of this integrated assessment approach is the focus of the data and modeling on the characterization of production systems – specific combinations of crop type, physical resources (soils, climate), capital and management. Site-specific data are used to characterize these production systems for operation of the Century model and for parameterization of the econometric production models for each system being considered. The Century model computes changes in soil carbon (C) and crop yield by system as a function of climate change and CO_2 change and the resultant feedbacks occurring within the crop-soil-water components of the ecosystem. These crop yield changes are incorporated into the econometric process simulation model, where the expected economic value of each system is simulated and the system with the highest expected value is selected for a given land unit and time period. Based on the system choice, the economic model simulates management decisions, production and net returns for each land unit and time period in the analysis. The information on system choice by land unit is combined with the Century model's data on soil C by system to calculate the soil C per land area. Finally, the data on soil C and net returns by land unit are aggregated and used to construct the spatial distributions of soil C and net returns for the population being studied.

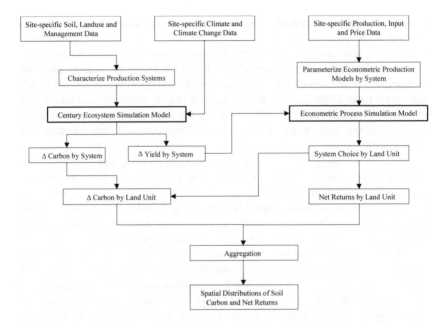

*Figure 5.1   Integrated assessment of climate change impacts in an
            agricultural production system using coupled Century and
            econometric process simulation models*

Integrated Assessment of Grain Production Systems in the US Northern Plains

In this section we describe the application of the integrated assessment approach to analyze the effects of adaptation and spatial heterogeneity on the impacts of climate change in the dryland grain production systems typical of the US northern plains region. We used two key variables as indicators of climate change impacts: soil organic carbon (SOC) as an indicator of ecosystem status, and net returns as an indicator of economic status.

For this analysis we utilized data from the principal agricultural zones of Montana (roughly the 100 000 square mile region east of the Rocky Mountains) where small grain production (wheat and barley) and pasture are the principal land uses. Some areas (such as the so-called Golden Triangle region of north-central Montana) are well-suited to the production of high-quality wheat and barley, whereas other areas are economically marginal for grain production due primarily to inadequate precipitation. Particularly in these marginal areas, relatively small reductions in precipitation could lead to substantial changes in production systems, primarily from grain production to pasture.

Several management strategies are used by farmers to deal with climate variability. A significant proportion of small grains are produced in a crop-fallow rotation, with the purpose of the fallow to accumulate soil moisture for subsequent crops. Many farmers follow the practice of fallowing a field every other year, but if soil moisture is adequate they may plant the field to a crop in two or more subsequent years. Yet other farmers (presumably in locations with greater precipitation) typically follow a continuous cropping strategy and rarely leave a field fallow.

Another management strategy to deal with climate variation utilizes the different growing periods of winter wheat (planted in the fall) and spring wheat (planted in the spring). For example, a farmer may plant a winter wheat crop in the fall, and then if the crop fails (because of adverse weather) a spring wheat crop may be planted in the same field.

Below we explain how the Century crop-ecosystem and econometric process models were characterized for the production systems in this region, followed by our analysis of the impacts of climate change.

The Century Crop-ecosystem Model

Century is a generalized biogeochemical ecosystem model that simulates carbon (biomass), nitrogen and other nutrient dynamics. The model has been used previously in several regional studies of climate change impacts on managed ecosystems (see for example Ojima et al. 1993; Parton et al. 1995; Paustian et al. 1996, 1997a). It includes sub-models for soil biogeochemistry, growth and yield sub-models for crop, grass, forest and savanna vegetation, and simple water and heat balance functions. The model employs a monthly time-step and the main input requirements include monthly precipitation and temperature, soil physical properties (such as texture and soil depth), atmospheric nitrogen inputs and management practices such as crop rotation, tillage and fertilization.

We used sub-divisions of MLRAs (Major Land Resource Areas) as the primary spatial units in our analysis (Figure 5.2). MLRAs are ecoregions defined by topographic and land use characteristics (SCS 1981). In addition to comprising areas of similar topography and land use, MLRAs are well-suited for broad scale analyses of agricultural systems because other databases such as soil maps are often georeferenced by MLRAs. In delineating spatial units, we first eliminated areas under forest cover, by overlaying land cover from Loveland et al. (1991). To further reduce within-region climate variability (particularly precipitation), we divided the MLRAs into lowland and plateau landform areas based on elevation data and climate conditions, yielding sub-regions labeled as 'low' (MLRA 52_l) and 'high' (MLRA 52_h) precipitation areas.

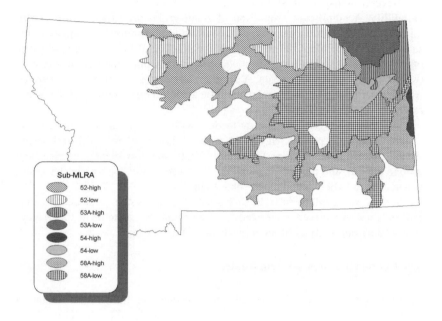

Figure 5.2   Major land resource areas in central and eastern Montana

Mean long-term climate variables for each polygon were calculated by overlaying the sub-MLRA coverage with the PRISM (Parameter-elevation Regressions on Independent Slopes Model; Daly et al. 1994) database, consisting of 4 km^2 polygons for the conterminous US. Climate variables in PRISM are based on 30-year mean (1960–1990) weather station records, spatially interpolated to account for topographic effects.

Soil physical properties for each polygon were derived from STATSGO soil association maps (SCS 1994) overlaid with the sub-MLRA coverage. For each STATSGO map unit occurring within a sub-MLRA, we extracted the component (soil series level, for example) attributes. For soils in Montana, the main characteristic needed for the Century model is surface soil texture. Soils that represented at least 5 percent of the area within the MLRA were included in the analysis and grouped according to soil texture classes, such that 3–5 textural classes were usually represented within each region. Areas within each textural class for map units occurring within a sub-MLRA were used for computing area-weighted averages of simulation results.

To initialize the model, we first ran the model for 8000 years to attain equilibrium values of soil C and N pools representative for native prairie, for each soil type within each spatial polygon (sub-MLRA). We then simulated two historical management scenarios for each soil–climate combination, involving either plow-out (in 1900) and cultivation under a wheat fallow rotation, or grazing on native rangeland. For the historical cropping sequence (1900–1975), data from long-term cropping experiments at Swift Current in southern Saskatchewan (Campbell and Zentner 1997) were used to parameterize the crop model for regional conditions and historic crop yields were calibrated to match with state and county average yields compiled by the National Agricultural Statistics Service (NASS 1999). To represent current conditions, the dominant management systems identified in the econometric model (spring cereal-fallow, continuous spring cereals, winter wheat-fallow, continuous winter wheat, cropland converted to perennial grasses and uncultivated rangeland) were run for 33 years (1975–1998) to represent present day (baseline) conditions.

The Econometric Process Model

The econometric process modeling approach was developed to simulate the farmer's choice of production system on a site-specific basis to assess economic and environmental impacts of changes in agricultural production systems (Antle and Capalbo 2001). In this approach, site-specific data are used to estimate econometric production models. These data and models are incorporated into a simulation model that represents the decision making process of the farmer as a sequence of discrete land use and continuous input use decisions on a site-specific basis. This modeling approach provides the capability to simulate the changes in the spatial distribution of agricultural production systems and changes in economic attributes of those systems (such as net returns) in response to changing climate scenarios.

The econometric process simulation model is constructed by estimating a system of behavioral equations (supply function and factor demand functions) for each production activity, and then carrying out a stochastic simulation of these equations using site-specific data. For each field, expected returns are computed and used to solve the discrete choice of production system on each land unit in each time period. Therefore the econometric process simulation model incorporates *endogenous adaptation* both in the form of changes in management decisions for a given system (as represented by its output supply and factor demand functions) and in the form of the choice among systems on each land unit.

Much of the economics literature on agricultural impacts of climate change is based on the use of representative-farm optimization models (see Kaiser et al. 1993; Darwin et al. 1995; Adams et al. 1999). The econometric process modeling approach differs fundamentally from an optimization model (a linear or non-linear programming model) in the way that farmer decision making is represented and simulated. In an econometric process model, economic decisions are based on draws from the *spatial and temporal distributions* of expected returns associated with each alternative land use or input choice associated with a production system. There is a positive probability that each feasible activity will be selected at each site. Thus, as repeated draws are made from the underlying statistical distributions, a realistic spatial and temporal distribution of competing activities is obtained. At the level of the population being represented in the simulations, corner solutions (solutions in which some choices do not occur) can only be obtained as a limiting case in which one activity economically dominates other activities at all sites or all time periods.

A limitation of the econometric process modeling approach, like other econometric models in the literature designed to represent the spatial variability in land-based production systems (see Pautsch et al. 2001; Plantinga et al. 1999; Stavins 1999), is the representation of producers as price takers, and thus the lack of a market adjustment mechanism. Impacts of price changes in the econometric process approach can be analyzed by using price scenarios derived from aggregate models or other sources.

In this application, the econometric models for winter wheat, spring wheat and barley crops were specified as a system of log-linear equations (a supply function, a machinery cost equation, and factor demands for variable fertilizer and pesticide inputs). These equations include the effects of fallow on supply and cost of production and thus capture the dynamics of the crop rotation being used. The equation systems were estimated using non-linear three-stage least squares with zero-degree homogeneity of the supply function and linear homogeneity of the variable cost function imposed.

To implement the econometric process simulation model, each of the 850 fields in the data set was described by total acres, location and an associated set of location-specific prices paid and received by the farmer. Sample distributions for type of tillage, use of crop insurance, and the previous crop were estimated from the survey data and were then sampled to initialize the model for each field. The econometric models were simulated to estimate expected output and cost of production and to calculate expected returns above short-run variable costs of production for each crop and management alternative on each field in each time period. To couple the econometric process model with the yields of the Century model, it was assumed that the supply functions for each crop are proportional to the percentage change in yield predicted by the Century model. Based on the maximization of expected returns, a fall decision is made to produce winter wheat, to produce a spring crop or fallow the field. If winter wheat is not grown, the model advances to the spring decision, where either spring wheat, barley, or fallow options are selected based on expected returns maximization.

The econometric process simulation model was calibrated to predict the observed mean frequencies of crops produced in the sample data. The model was calibrated using three parameters: the expected yield variability, the discount rate, and the expected future crop price. To validate the model, both within-sample and out-of-sample tests were performed (for details see Antle and Capalbo 2001).

Analysis of Climate Change Scenarios

We analyzed impacts for different combinations of changes in climate and atmospheric CO_2. Baseline (present) climate was derived from the PRISM dataset as described above in reference to the crop-ecosystem model. For climate under a 2x CO_2 scenario we used results from equilibrium simulations of the Canadian Climate Change (CCC) general circulation model (GCM), compiled in the VEMAP dataset (VEMAP 1995). The CCC model uses 0.5×0.5 degree grid cells and there are only a few cells spanning the Montana study area. Therefore, we used the average latitude and longitude of each sub-MLRA to determine the nearest grid cell used in the CCC model simulation. The simulated monthly changes in temperature (expressed as a difference) and precipitation (expressed as a ratio) of projected climate, relative to current climate from the VEMAP dataset, were applied to our base climate (PRISM-derived) for each sub-MLRA. Under the 2x CO_2 conditions, changes in MAP (mean annual precipitation) ranged between 2–5 cm across the sub-MLRAs and the increase in MAT (mean annual temperature) was between 4.5 and 4.8°C (Figure 5.3). These estimates for changes in MAP and MAT are consistent with the range of estimates provided by the IPCC Third Assessment Report (IPCC 2001). Carbon dioxide levels were set at current ambient level (370 ppm) or 640 ppm, which corresponds to CO_2 levels for the atmospheric composition under the 2x CO_2 climate scenario.

To generate input to the econometric model for climate change and CO_2 change impacts on productivity and soil carbon, we simulated all possible management transitions for each of the six production systems, yielding a total of 36 simulations for each soil-by-sub-MLRA combination, for each climate change scenario. The scenarios were simulated by imposing a step change in climate and/or CO_2 and then simulating the model for a 50-year period and computing the percentage change in crop productivity and soil C (after 50 years), relative to the baseline condition. Thus, the model results represent a quasi-static equilibrium analysis, in that we did not simulate transient dynamics in either climate or atmospheric CO_2.

The relative changes in productivity from Century were incorporated into the econometric simulation model, which then simulated the changes in management according to the economic criteria that determine the farmer's decision making. Thus, changes in management (adaptation) in response to

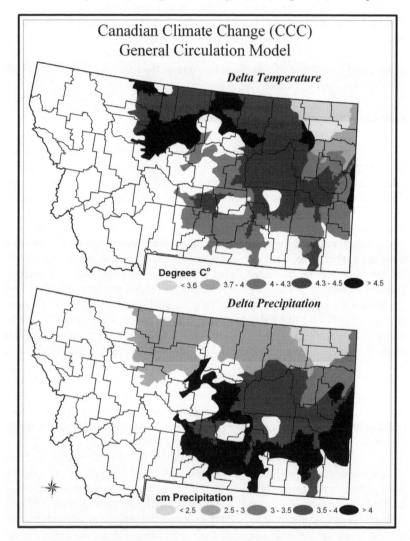

Notes: Values are changes relative to 30-year (1960–1990) mean climate; both temperature and precipitation increase for the region under the climate change scenario.

Figure 5.3   *Projected changes in mean annual temperature (MAT) and mean annual precipitation (MAP) from the Canadian Climate Center model for the study region under 2x CO$_2$ conditions*

changes in climate is an endogenous property of the two models working together. That is, adaptation responses are not specified as *a priori* conditions or scenarios, but rather derive internally from the analysis. In several other studies of climate change impacts using crop simulation models (see for example Easterling et al. 1993, 2001; Rosenzweig et al. 1994), exogenous agronomic adaptations were incorporated or used as the basis for determining the scenarios, such as changes in assumed planting and/or harvest dates. Most of these crop simulation models operate on a daily time-step, and crop growth can be sensitive to relatively small changes in assumed timing of planting and other management operations. The Century model, in contrast, is designed to simulate response of the crop-ecosystem over long time periods and operates on a monthly time-step, and thus is less sensitive to specific assumptions about planting date and timing of other management decisions. Since our analysis focuses on endogenous economic adaptations brought about through choice of production system, we do not impose exogenous adaptation assumptions through manipulation of the ecosystem model.

RESULTS

The yield changes simulated by Century (by crop, production system and sub-MLRA) are shown in Figures 5.4, 5.5, and 5.6 for alternative climate and CO_2 scenarios. These data indicate that changes in climate without the effects of CO_2 fertilization resulted in a decline in grain yields for all systems (indicated in the figures as follows: grass, WW = winter wheat-continuous, WWF = winter wheat-fallow, SW = spring wheat-continuous, SWF = spring wheat-fallow) and all sub-MLRAs, with yield levels ranging from 45 percent to 80 percent of baseline values. In contrast, grass yields increased by 10 percent to 20 percent in response to changes in climate, mainly due to the increase in temperature (Figure 5.4). Yields of all crops increased under elevated carbon dioxide levels with winter and spring wheat showing positive responses of 17 percent to 55 percent across the sub-MLRAs (Figure 5.5). The crops grown on fields that were previously fallowed often showed less of an increase than the same crop grown on land that was previously cropped. When the effects of climate and elevated CO_2 are combined, the changes in yields tended to be offsetting (Figure 5.6). The net result was an overall decline in spring wheat yields of 20–30 percent, and an increase in winter wheat yields and yields of grass grown for pasture. Looking at spatial variability in yields, winter wheat yields increased in nearly all sub-MLRAs, while spring wheat yields declined in all areas except 53A-high and 53A-low, which are the colder temperature areas.

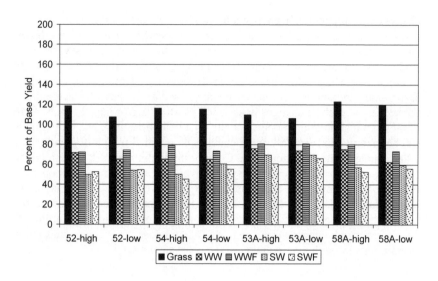

Figure 5.4   *Yield changes predicted by the Century model for climate change scenario (temperature and precipitation changes), by Montana sub-MLRA.*

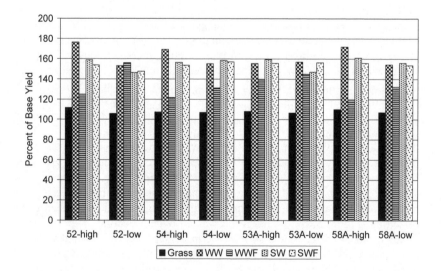

Figure 5.5   *Yield changes predicted by the Century model for CO_2 fertilization scenario, by Montana sub-MLRA.*

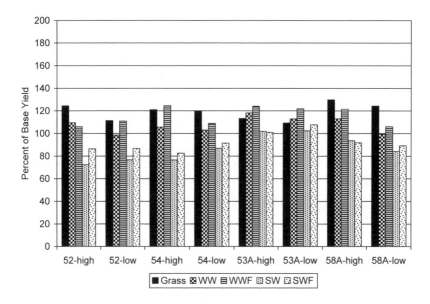

Figure 5.6   *Yield changes predicted by the Century model for climate change + CO₂ scenario, by Montana sub-MLRA.*

Expected Returns and Choice of Production System

The econometric process model computes expected net returns by production system to determine the choice of system. Figures 5.7, 5.8 and 5.9 present results for mean net returns by sub-MLRA and by climate change and CO_2 scenario. Figures 5.10 and 5.11 present system choice results graphically for sub-MLRAs 52-high and 58A-low to illustrate the types of changes simulated by the model. The simulations represent four systems: winter wheat in a rotation with fallow (WWF); spring wheat and barley in rotation with fallow (SWBF); spring wheat and barley in continuous rotations (SWB); and grass for pasture or land conservation (continuous winter wheat is also included as an option in the model, but is little used in either base or climate change simulations, so it is not included in the results presented here). The base scenario replicates observed data showing that about one-third of agricultural land is in fallow and grass. About two-thirds of the remainder is allocated to crop-fallow systems.

Figure 5.7 shows that base-climate net returns vary substantially across the sub-MLRAs, ranging from over $100 per acre in sub-MLRAs 52-high and 52-low (known to be the most productive grain producing region in Montana), to $60–$70 per acre in the other sub-MLRAs (note, these net returns are revenue minus variable costs of production, and can be interpreted as returns to land and capital ownership, family labor, and risk). Figures 5.7, 5.8 and 5.9 show that net returns tend to follow the patterns of yield changes shown in Figures 5.4, 5.5 and 5.6, and reflect the relative advantage of each sub-MLRA in the different systems. The climate-change-only scenario shows a substantial negative impact on returns in all sub-MLRAs, with the largest absolute impact in the most productive areas where the highest proportion of land is in crops. The CO_2-only scenario shows, conversely, a highly positive impact on returns associated with the higher yields, and the combined CC+CO_2 scenario shows that the negative impacts of changes in temperature and precipitation in the CC scenario are offset by the positive effects of CO_2 fertilization. Five of the eight areas show a net positive impact on net returns under the CC+CO_2 scenario relative to the base climate scenario.

Compared to the no-adaptation scenarios, all scenarios (Figures 5.7, 5.8 and 5.9) show a smaller negative impact of CC with adaptation, and a larger positive impact of CO_2 fertilization, as expected. Figures 5.10 and 5.11 show that with the CC scenario, the negative impacts on yields and returns to spring wheat and barley relative to winter wheat and grass cause production system utilization to shift significantly towards these latter two systems. Thus the results support the hypothesis that climate change in the northern Great Plains region would tend to shift utilization of production systems towards the pattern now typical of the central Great Plains where winter wheat predominates crop production. The increases in temperature in the CC-only scenario also cause significant shifts from crops towards grass in lower-productivity regions, presumably because any moisture increases are not adequate to sustain crop production at the higher temperatures in areas that are already marginal with respect to precipitation under the base climate. These patterns are illustrated by the results for sub-MLRAs 52-high and 58A-low (Figures 5.10 and 5.11).

Climate Change Impacts on Soil Carbon

Soil organic matter, which contains about 55 percent carbon by mass, plays several key roles in the function of agro-ecosystems and as such is a widely used indicator of soil quality and the overall sustainability of agricultural ecosystems (Doran 2002). Soil organic matter has a direct link to soil fertility and crop productivity as a major store of essential plant nutrients, such as nitrogen and phosphorus, and it also provides buffer capacity for other nutrients such as calcium and magnesium, due to its high cation exchange capacity. Equally important, soil organic matter helps to promote soil aggregation and favorable soil structure, hence increasing water infiltration, reducing surface runoff and increasing soil water storage. Historically, most of the cropland soils in the Northern Great Plains, particularly with summer

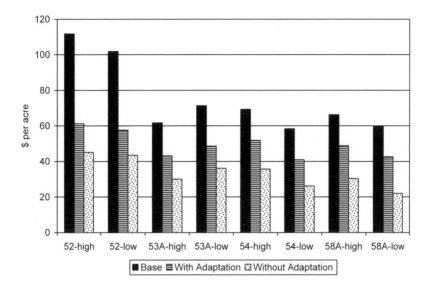

Figure 5.7   *Mean net returns for climate change scenario, by Montana sub-MLRA*

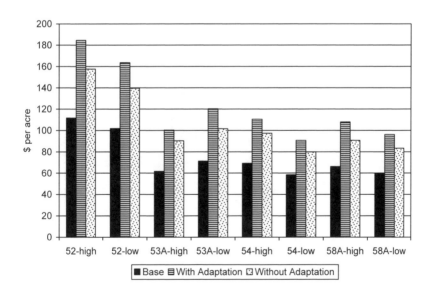

Figure 5.8   *Mean net returns for CO₂ scenario, by Montana sub-MLRA*

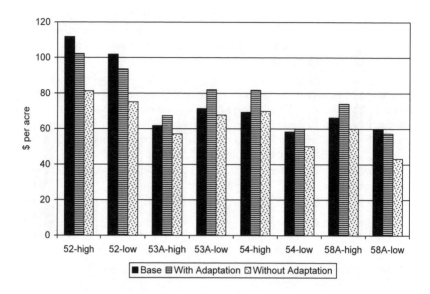

*Figure 5.9   Mean net returns for climate change + CO₂ scenario, by Montana
                sub-MLRA*

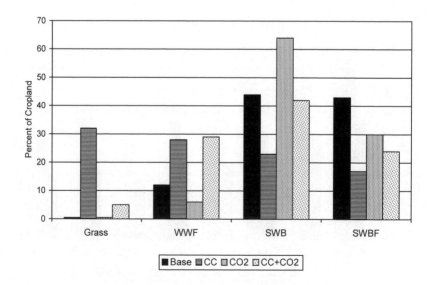

*Figure 5.10   Production system utilization for sub-MLRA 52-high under
                climate change and CO₂ scenarios*

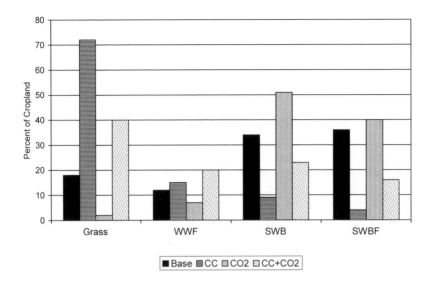

Figure 5.11   *Production system utilization for sub-MLRA 58A-low under climate change and CO_2 scenarios*

fallow, have lost a substantial portion of their original soil organic carbon stocks, on the order of 30–60 percent in the top soil (Paustian et al. 1997b). Hence, maintenance and restoration of soil carbon stocks has become an important management concern in the region (Peterson et al. 1998).

Changes in soil carbon contents are governed by the balance between carbon added to the soil, largely through crop residues and roots, and the rate of decomposition (and release of carbon as CO_2) of organic matter already in the soil. Environmental conditions, particularly soil temperature and moisture, influence both decomposition rates as well as crop productivity (thereby influencing crop residue addition rates). Agricultural management practices, such as crop choice, rotation, tillage, fertilization and other activities, directly affect productivity and carbon inputs; most management practices also influence decomposition rates via crop influences on temperature, moisture and nutrient availability, residue decomposability and tillage-induced soil disturbance (Paustian et al. 2000).

Accordingly changes in soil carbon stocks due to the climate change scenarios, as well as the adaptive responses of farmers, are a result of the interaction of multiple influencing factors and feedbacks within the agro-ecosystem. However, several distinct patterns in soil carbon responses emerged from our analysis and their interpretation suggests important differences between cropping systems and sub-regions, and in the relative importance of climatic influences and management practices in controlling soil C stocks. The impact of climate change on ecosystem properties, where the present mix of land use and management practices remains constant (without adaptation), will be discussed first, followed by a discussion of ecosystem impacts allowing for endogenous adaptation by farmers in response to climate and CO_2 changes.

Impacts on Carbon Without Adaptation

In most cases, climate change alone tended to decrease soil carbon stocks, due to a reduction in biomass productivity as well as increased decomposition rates, mainly due to higher temperatures (Figure 5.12). Exceptions to this were winter wheat-fallow systems and the perennial grass systems (illustrated for the grass-hay system in Figure 5.12). For winter wheat-fallow, yield declines were primarily due to a decrease in the harvest index (the ratio of harvested yield to above ground biomass), which in turn was the result of higher moisture stress. Thus, while simulated grain yields in winter wheat-fallow were reduced by 20–25 percent under climate change (Figure 5.4), total above ground biomass was between 90–106 percent (data not shown) of that under ambient climate and hence, on average, residue input rates were slightly increased under climate change. These higher residue inputs tended to offset the effects of higher decomposition rates, resulting in little change in soil organic carbon stocks. For perennial grass crops such as hay, higher temperatures along with the increased precipitation made for longer growing seasons and increased productivity, and hence substantially increased carbon inputs from root turnover.

Soil carbon levels in all of the cropping systems responded positively to doubled CO_2 levels, although the largest responses were in the continuous grain systems and with perennial grass (Figure 5.12). Increased CO_2 can directly increase photosynthetic rates in wheat and other crops and also increases plant water use efficiency (Acock and Allen 1985); both mechanisms are incorporated in the Century model. The higher relative increase in biomass productivity and the greater increase in soil carbon in continuous versus fallow cropping systems suggests that the predominant effect of CO_2 was to increase water use efficiency (which will have a greater impact under continuous cropping due to higher water demand).

In the scenario with both climate change and doubled CO_2, the effects of increased CO_2 tended to offset some of the negative impacts of climate change alone or gave an additional positive response in soil carbon contents (for the winter wheat-fallow and perennial grass systems). However, the effect of climate change and CO_2 together were not strictly additive. For the annual cropping systems, the overall impacts on soil carbon stocks were relatively

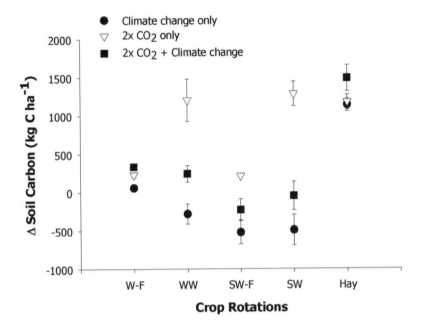

Notes: Management systems are: W-F=winter wheat-fallow; WW=continuous winter wheat; SW-F=spring wheat-fallow; SW=continuous spring wheat; Hay=perennial grass harvested for hay). Point estimates represent mean changes (kg C/ha) from the initial baseline condition, averaged over all sub-regions and bars represent the variability (as 1 SD) among sub-regions.

Figure 5.12   *Simulated changes in soil carbon stocks (0–20 cm depth) after 50 years following a step change in climate and/or CO₂, for the major agricultural systems, with no changes in management.*

minor (averaging between -230 kg/ha and +340 kg/ha), or less than a 10 percent change in total stocks simulated under baseline conditions.

There were differences between sub-regions under the climate and CO_2 scenarios and the highest variability in soil C response tended to occur under the climate change only scenario (Figure 5.12). Variability in absolute rates of soil carbon change was related, in part, to geographic differences in total stocks of soil C. For example, in the cereal-fallow systems, under climate change and ambient CO_2, sub-regions having the highest total C stocks (usually those in the higher precipitation zones) tended to have the highest loss rates. Thus the relative rates of soil C loss were more similar across sub-regions. In contrast, for the continuous cropping systems, the sub-regions with

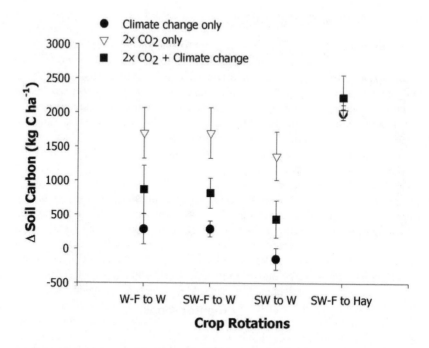

Notes: Management systems are: W-F=winter wheat-fallow; W=continuous winter wheat; SW-F=spring wheat-fallow; SW=continuous spring wheat; Hay=perennial grass harvested for hay). Point estimates represent mean changes (kg C/ha) from the initial baseline condition, averaged over all sub-regions and bars represent the variability (as 1 SD) among sub-regions.

Figure 5.13   Simulated changes in soil carbon stocks (0–20 cm depth) after 50 years following a step change in climate and/or CO₂, for selected management changes following the climate perturbation.

more favorable climate conditions (for example high versus low precipitation sub-regions) and the highest initial soil carbon levels tended to have lower losses (or higher gains) for climate change and doubled CO_2 scenarios. This contrasting behavior between the fallow and continuous cropping systems suggests that in the systems with fallow, climate change impacts on decomposition rates were the main control on the soil carbon balance, while for the continuous cropping systems the impacts on crop productivity and residue inputs played a relatively greater role.

Impacts on Carbon With Adaptation

As noted above, the dominant trends in land allocation changes predicted by the economic model under the climate change scenarios were towards more winter wheat and either more continuous cropping rotations (doubled CO_2) or more grass (climate change plus doubled CO_2). In general these trends in management changes resulted in significantly higher soil carbon stocks as compared to the results for the scenarios when cropping choices remain static (no adaptation). Examples of average soil carbon stock changes for some of the dominant system transitions are shown in Figure 5.13. With no adaptation, most annual cropping systems are simulated as losing soil carbon with climate change alone. However, with adaptation the most significant management change was conversion of annual crop systems to grass (Figures 5.10 and 5.11), which had the highest rates of soil carbon gain under all climate scenarios. Across sub-regions, for all climate and CO_2 scenarios, the amount of soil C increase following management transitions to continuous cropping tended to be highest for the higher precipitation zones, which showed the largest relative gains in productivity. Again this is indicative that the key factor determining soil carbon levels, when allowing for adaptive responses, is the impact of increased residue inputs.

In summary, the spatial distribution of climate change impacts on soil C under each of the three scenarios, with and without management, shows that adaptation has substantial effects in ameliorating negative changes and strengthening positive changes in soil C. The results suggest that land use and management practices exert a relatively strong control on soil organic carbon dynamics. Further, the analysis suggests that farmer adaptation to climate change, driven by economic decisions regarding changes in land use and input use in order to maximize net returns, is consistent with reducing potentially negative effects of climate change on soil organic matter stocks and soil sustainability.

Analysis of Economic Vulnerability to Climate Change

Antle et al. (2004) present a conceptual model for economic analysis of adaptation to climate change and use this model to define several measures of vulnerability. In this model, economic agents (farmers) make land use and management decisions represented by the vector x to maximize economic value. In the analysis presented here, farmers make land use and management decisions to maximize economic returns $v(x, z, c, p, w)$, where c represents the climate regime, z is fixed capital investment, p is crop price, and w is input price. Thus, for the climate regime c_0 the optimal choice is $x_0 = x(z_0, c_0, p_0, w_0)$ and economic returns are $v(x_0, z_0, c_0, p_0, w_0)$. When there is a change in climate to regime c_1, x_0 becomes economically sub-optimal, yielding returns $v(x_0, z_0, c_1, p_0, w_0) < v(x_0, z_0, c_0, p_0, w_0)$. Accordingly, Antle et al. (2004) define the *relative climate vulnerability – no adaptation* of the system as the maximum potential loss relative to the base climate,

$$RCVN = [v(x_0, z_0, c_0, p_0, w_0) - v(x_0, z_0, c_1, p_0, w_0)]/\ v(x_0, z_0, c_0, p_0, w_0).$$

Adapting management optimally to the climate regime c_1 results in a decision x_1 and economic returns to the system of $v(x_1, z_0, c_1, p_0, w_0) > v(x_0, z_0, c_1, p_0, w_0)$. Antle et al. (2004) define the loss associated with climate change and optimal adaptation as the *relative climate vulnerability – adaptation* of the system, and measure it as

$$RCVA = [v(x_0, z_0, c_0, p_0, w_0) - v(x_1, z_0, c_1, p_0, w_0)]/\ v(x_0, z_0, c_0, p_0, w_0).$$

Therefore a manager who does adapt to the changed climate conditions gains the amount $v(x_1, z_0, c_1, p_0, w_0) - v(x_0, z_0, c_1, p_0, w_0)$ relative to the manager who does not adapt, and the *relative adaptive gain* of the system is accordingly defined as

$$RAG = [v(x_1, z_0, c_1, p_0, w_0) - v(x_0, z_0, c_1, p_0, w_0)]/\ v(x_0, z_0, c_0, p_0, w_0).$$

Note that vulnerabilities are positive when there is a loss, and negative where there is a net gain from climate change. These concepts can be translated into absolute rather than relative changes by multiplying them by the base value $v(x_0, z_0, c_0, p_0, w_0)$.

Vulnerability also can be measured in relation to a threshold value (IPCC 1996; Schimmelpfennig and Yohe 1999). Let an absolute threshold for the value function be a fixed value AT defined in the units of the value function, and let a relative threshold be $0 < RT < 1$. Antle et al. (2004) define absolute or relative vulnerability, without and with adaptation, as the probability that the value function under a perturbed climate (c_1) would be less than or equal to the absolute or relative threshold:

$$ATCVN = Pr\{v(x_0, z_0, c_1, p_0, w_0) < AT\}$$
$$ATCVA = Pr\{v(x_1, z_0, c_1, p_0, w_0) < AT\}$$
$$RTCVN = Pr\{v(x_0, z_0, c_1, p_0, w_0) < RT\cdot v(x_0, z_0, c_0, p_0, w_0)\}$$
$$RTCVA = Pr\{v(x_1, z_0, c_1, p_0, w_0) < RT\cdot v(x_0, z_0, c_0, p_0, w_0)\}.$$

Systems with relatively poor endowments of capital (human, social, physical and natural) have been hypothesized to be less able to adapt to climate change, and thus more vulnerable, than systems with relatively better endowments (IPCC 2001). The analytical framework outlined here can be used to investigate this hypothesis.

Figures 5.14, 5.15, and 5.16 present the mean values of RCVN, RCVA and RAG by sub-MLRA, plotted against mean net returns from the base scenario (a measure of resource endowment). Note that the measures of vulnerability are defined for given values of prices. To investigate the sensitivity of the measured vulnerability to changes in prices, the results are presented for three combinations of relative output prices: high grain prices and low grass prices; medium prices for each (corresponding to the prices used in the results reported above); and low grain prices and high grass prices. As we noted earlier, sub-MLRAs 52-high and 52-low have the best endowments and consequently earn the highest mean net returns in the base case (and in all climate scenarios).

Figure 5.14 shows the results for relative climate vulnerability – no adaptation (RCVN). In the case of medium wheat and grass prices, the vulnerability of sub-MLRAs 52-high and 52-low is about 60 percent (these are the two points corresponding to a mean net returns value near 100). The other six sub-MLRAs correspond to the cluster of data points with lower net returns, and their vulnerability is in the range of 0.47 to 0.65. Thus, with this set of output prices, there is no clear relationship between resource endowments and vulnerability. The data points towards high grain prices, and low grass prices likewise show no evidence of a negative relationship – if anything, the high grain prices cause the more productive regions to be more vulnerable to losses associated with climate change. However, in the case of low grain prices and high grass prices, there is a negative relationship. This negative relationship reflects the greater utilization of the grass-based production system in the regions with poorer endowments. Without adaptation, these areas are much more vulnerable to reductions in grass yields when grass prices are high.

Figure 5.15 shows the measures of relative climate vulnerability with adaptation (RCVA). Again, the results do not support a negative relationship between vulnerability and resource endowments. In the case of low grain prices and high grass prices, the regions with poor resource endowments actually have negative vulnerability, that is, they gain from climate change, due to their ability to shift to grass-intensive systems. In contrast, with adaptation, the regions with better resource endowments (particularly sub-MLRAs 52-high and 52-low) are vulnerable to climate change (with RCVA about 20 percent) because of their greater reliance on grain production.

Figure 5.16 shows the measures of relative adaptive gain (RAG) associated with the three output prices scenarios. These results show that the regions with poorer resource endowments tend to gain more from adaptation than the regions with better resource endowments.

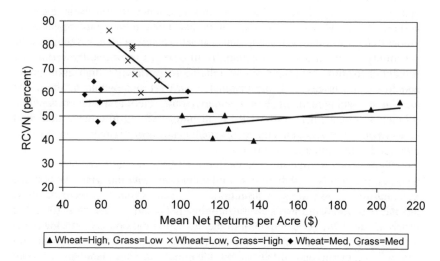

Figure 5.14   Relative mean climate vulnerability – no adaptation (RCVN),
by Montana sub-MLRA, for low, medium and high price
scenario

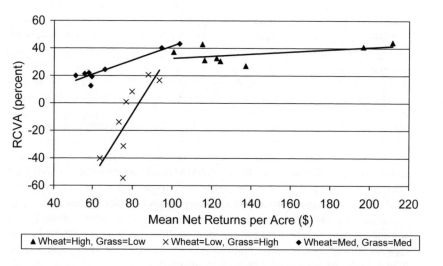

Figure 5.15   Relative mean climate vulnerability – adaptation (RCVA), by
Montana sub-MLRA, for low, medium, and high price scenario

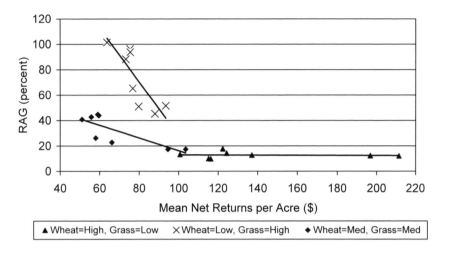

Figure 5.16   Relative mean climate vulnerability adaptive gain (RAG), by Montana sub-MLRA, for low, medium and high price scenario

As noted above, it is also possible to construct a corresponding set of absolute measures of vulnerability. Although these results are not presented here, the same general conclusions hold for the absolute measures; in particular, there is no necessary negative relationship between vulnerability and resource endowments.

Figures 5.17 and 5.18 present the results for the absolute and relative threshold-based measures of vulnerability. These results show that the relative threshold measures of vulnerability show no systematic relationship to resource endowments across regions. However the absolute threshold measures of vulnerability do show a consistent negative relationship to resource endowments. The explanation for this result seems evident: well-endowed regions may be impacted in percentage terms as much or more than poorly endowed regions. However poorly endowed regions are closer to any absolute threshold than better-endowed regions, and therefore are at higher risk of falling below an absolute threshold when subjected to a climate change that has a generally adverse effect on productivity.

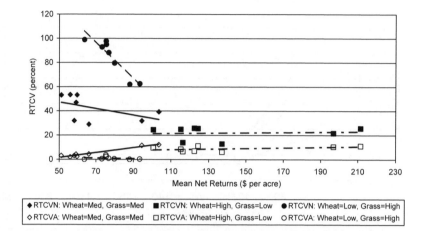

Figure 5.17   Relative threshold climate vulnerability (RTCV), with (A) and without (N) adaptation, for low, medium and high price scenario

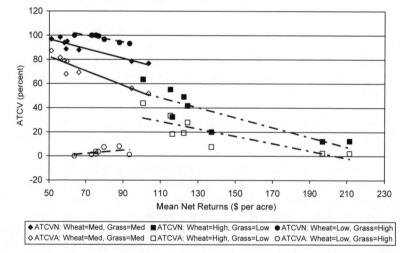

Figure 5.18   Absolute threshold climate vulnerability (ATCV), with (A) and without (N) adaptation, for low, medium and high price scenario

The results in Figures 5.14–5.18 show that economic conditions may have a substantial impact on the degree of vulnerability and furthermore interact with vulnerability and adaptation in complex ways. Figures 5.17 and 5.18 show that vulnerability, however measured, decreases with the degree of adaptation. These figures also show that the degree to which adaptation affects vulnerability interacts strongly with crop prices. The case of low grain prices and high grass prices shows the particularly extreme outcome wherein with adaptation all regions show very low vulnerability, whereas without adaptation the regions show a very high degree of vulnerability that is strongly correlated with resource endowments.

CONCLUSIONS

In this chapter we describe an integrated modeling approach that exhibits two key features needed to study the response of agricultural production systems to climate change: it represents adaptation as an endogenous economic response to climate change; and it provides the capability to represent the spatial variability in biophysical and economic conditions that interact with adaptive responses. We apply this approach to the dryland grain production systems of the Northern Plains region of the US to investigate the ecological and economic impacts of climate change in this agricultural system.

Our findings show that the spatial distributions of net returns, and thus the choice of production system, vary systematically with assumptions about climate impacts, CO_2 fertilization and adaptation. Accordingly the ecological impacts of climate change also vary according to these factors. We find that adaptation has substantial effects on both ecological outcomes (changes in soil carbon) and economic vulnerability. Without adaptation each system is characterized by lower soil carbon stocks, as well as lower mean economic returns and higher spatial variability in returns, relative to the same system simulated with endogenous adaptation.

Our results show that the vulnerability of agriculture to climate change depends on how it is measured (in relative versus absolute terms, and with respect to a threshold), and on the complex interactions between climate change, CO_2 level, adaptation and economic conditions. When we subject our coupled models to a sensitivity analysis of economic assumptions, we find that the degree of vulnerability of wealthier and poorer regions can change substantially. Our results show that relative vulnerability does not generally increase as resource endowments become poorer; indeed, we find that without adaptation, there may be either a positive or negative association between endowments and relative vulnerability. However, vulnerability measured in relation to an absolute threshold does vary inversely with resource endowments. Measures of vulnerability are found to interact strongly with assumed economic conditions and the degree of adaptation. Finally, we find that there is generally a positive relationship between gains from adaptation and the resource endowment of a region. This finding underlines the particularly important role that adaptation plays in mitigating climate change impacts in poorer regions.

The results of our analyses suggest that the studies based on representative farm models, based primarily on representative data from well-endowed regions, most likely fail to measure the impacts of climate change on farms that are located in areas with less favorable resource endowments. Our findings validate concerns expressed in the IPCC Third Assessment Report (2001) about the validity of the results reported in the literature based on large-scale regional and national studies, wherein representative farm models are used to extrapolate impacts to larger regions.

We conclude that the distribution of the economic impacts of climate change across regions with more and less favorable resource endowments is likely to depend on the ability of farmers in those regions to adapt to climate change. This finding has particularly significant implications for the distribution of impacts within developing countries, as well as between the more and less developed countries. It provides empirical evidence to support the hypothesis advanced in the IPCC Second and Third Assessment Reports (1996, 2001) that climate change is likely to have its greatest adverse impacts on areas where resource endowments are the poorest and the ability of farmers to respond and adapt is most limited.

This methodology can also be used to undertake sensitivity analysis of the key assumptions and parameters in the econometric process model and the biophysical model. Sensitivity analysis would provide changes in the distributions of net returns and land use patterns in a given region as well as changes in the economic welfare of producers in response to changes in yields, economic variables such as prices and input costs, and management techniques. In addition, sensitivity analyses could be used to begin to address the issue of how uncertainty in the parameters of process-based models (for which we lack statistical distributions) affects the results of an integrated assessment simulation. An example of this type of analysis can be found in Antle et al. (2002).

REFERENCES

Acock, B. and L.H. Allen, Jr (1985), 'Crop responses to elevated carbon dioxide concentrations', in B.R. Strain and J.D. Cure (eds), *Direct Effects of Increasing Carbon Dioxide on Vegetation* (DOE/ERB0238), Washington, DC: US Department of Energy, pp. 53–97.
Adams, R.M., B.A. McCarl, K. Segerson, C. Rosenzweig, K.J. Bryant, B.L. Dixon, R. Conner, R.E. Evenson and D. Ojima (1999), 'The economic effects of climate change on US agriculture', in R. Mendelsohn and J.E.

Neumann (eds), *The Impact of Climate Change on the United States Economy*, Cambridge, UK: Cambridge University Press, pp. 18–54.

Antle, J.M. and S.M. Capalbo (2001), 'Econometric process models for integrated assessment of agricultural production systems', *American Journal of Agricultural Economics*, **83**(2): 389–401.

Antle, J.M. and J.J. Stoorvogel (2001), 'Integrating site-specific biophysical and economic models to assess tradeoffs in sustainable land use and soil quality', in N. Heerink, H. Van Keulen and M. Kuiper (eds), *Economic Policy and Sustainable Land Use: Recent Advances in Quantitative Analysis for Developing Countries*, New York: Physica-Verlag, pp. 169–84.

Antle, J. M., S. M. Capalbo, S. Mooney, E. Elliott, and K. Paustian (2002), 'Sensitivity of carbon sequestration costs to soil carbon rates', *Environmental Pollution*, **116**(3): 413–22.

Antle, J.M., S.M. Capalbo, E.T. Elliott, and K.H. Paustian (2004), 'Adaptation, spatial heterogeneity, and the vulnerability of agricultural systems to climate change and CO_2 fertilization: an integrated assessment approach', *Climatic Change*, **64**(3): 289–315.

Campbell, C.A. and R.P. Zentner (1997), 'Crop production and soil organic matter in long-term crop rotations in the semi-arid Northern Great Plains of Canada', in E.A. Paul, E.T. Elliott and C.V. Cole (eds), *Soil Organic Matter in Temperate Ecosystems: Long-Term Experiments in North America*, Boca Raton, FL: Lewis Publishers, CRC Press, pp. 317–33.

Daly, C., R.P. Neilson, and D.L. Phillips (1994), 'A statistical-topographic model for mapping climatological precipitation over mountainous terrain', Terrain, *Journal of Applied Meteorology*, **33**: 140–58.

Darwin, R., M. Tsigas, J. Lewandrowski and A. Raneses (1995), *World Agriculture and Climate Change: Economic Adaptations*, Natural Resources and Environmental Division, Economic Research Service, US Department of Agriculture, Agricultural Economic Report No. 703.

Doran, J.W. (2002), 'Soil health and global sustainability: translating science into practice', *Agriculture Ecosystems and Environment*, **88**: 119–27.

Easterling, W.E., P.R. Crosson, N.J. Rosenberg, M.S. McKenny, L.A. Katz and K.M. Lemon (1993), 'Agricultural impacts of and responses to climate change in the Missouri-Iowa-Nebraska-Kansas (MINK) Region', *Climate Change*, **24**(1/2): 23–61.

Easterling, W.E., L.O. Mearns, C.J. Hays and D. Marx (2001), 'Comparison of agricultural impacts of climate change calculated from high and low resolution climate change scenarios: part II: accounting for adaptation and CO_2 direct effects', *Climate Change*, **51**(2): 173–97.

Hansen, J.W. and J.W. Jones (2000), 'Scaling-up crop models for climate variability applications', *Agricultural Systems*, **65**(1): 43–72.

Intergovernmental Panel on Climate Change (IPCC) (1996), *Climate Change 1995: The Science of Climate Change, Contribution of Working Group I to the Second Assessment Report of the Intergovernmental Panel on Climate Change*, J.T. Houghton, L.G. Meira Filho, B.A. Callander, N. Harris, A. Kattenberg and K. Maskell (eds), World Meteorological Organization, and the United Nations Environment Programme, Cambridge, UK: Cambridge University Press, 572 pp.

Intergovernmental Panel on Climate Change (IPCC) (2001), *Climate Change 2001: Impacts, Adaptation, and Vulnerability, Contribution of Working Group II to the Third Assessment Report of the Intergovernmental Panel on Climate Change*, J.J. McCarthy, O.F. Canziani, N.A. Leary, D.J. Dokken and K.S. White (eds), World Meteorological Organization and the United Nations Environment Programme, Cambridge, UK: Cambridge University Press, 1032 pp.

Kaiser, H.M. (1999), 'Assessing research on the impacts of climate change on agriculture', in G. Frisvold and B. Kuhn (eds), *Global Environmental Change and Agriculture: Assessing the Impacts*, Cheltenham, UK: Edward Elgar Publishing, pp. 221–38.

Kaiser, H.M., S.J. Riha, D.S. Wilks, D.G. Rossiter and R. Sampath (1993), 'A farm-level analysis of economic and agronomic impacts of gradual climate warming', *American Journal of Agricultural Economics*, **75**(2): 387–98.

Lewandrowski, J. and D. Schimmelpfennig (1999), 'Economic implications of climate change for US agriculture: Assessing recent evidence', *Land Economics*, **75**(1): 39–57.

Loveland, T.R., J.W. Merchant, D.O. Ohlen and J. F. Brown (1991), 'Development of a land cover characteristics database for the conterminous United States', *Photogrammetric Engineering and Remote Sensing*, **57**(11): 1453–63.

Mendelsohn, R., W.D. Nordhaus and D. Shaw (1994), 'The impact of global warming on agriculture: a ricardian analysis', *American Economic Review*, **84**(4): 753–71.

Metherell, A.K., L.A. Harding, C.V. Cole and W.J. Parton (1993), 'CENTURY Soil Organic Matter Model Environment, Agroecosystem version 4.0', Great Plains System Research Unit Technical Report No. 4, USDA-ARS, Fort Collins, CO.

National Agricultural Statistics Service (NASS) (1999), Published Estimates Database, USDA-National Agricultural Statistics Service, data available online at http://www.nass.usda.gov:81/ipedb/.

Ojima, D.S., W.J. Parton, D.S. Schimel, J.M.O. Scurlock and T.G.F. Kittel (1993), 'Modeling the effects of climatic and CO_2 changes on grassland storage of soil C', *Water, Air and Soil Pollution*, **70**(1–4): 643–657.

Parton, W.J., D.S. Schimel, D.S. Ojima, and C.V. Cole (1994), 'A general model for soil organic matter dynamics: sensitivity to litter chemistry, texture and management', in R.B. Bryant and R.W. Arnold (eds), *Quantitative Modeling of Soil Forming Processes*, SSSA Special Publication Number 39, Madison, WI: Soil Science Society of America, pp. 147–167.

Parton, W.J., J.M.O. Scurlock, D.S. Ojima, D.S. Schimel, D.O. Hall, M.B. Coughenour, E. Garcia Moya, T.G. Gilmanov, A. Kamnalrut, J.L. Kinyamario, T. Kirchner, S.P. Long, J-C. Menaut, O.E. Sala, R.J. Scholes and J.A. van Veen (1995), 'Impact of climate change on grassland production and soil carbon worldwide', *Global Change Biology*, 1: 13–22.

Paustian, K., E.T. Elliott, G.A. Peterson and K. Killian (1996), 'Modeling climate, CO_2 and management impacts on soil carbon in semi-arid agroecosystems', *Plant Soil*, **187**(2): 351–65.

Paustian, K., E.T. Elliott and K. Killian (1997a), 'Modeling soil carbon in relation to management and climate change in some agroecosystems in Central North America', in R. Lal, J.M. Kimble, R.F. Follett and B.A Stewart, (eds), *Soil Processes and the Carbon Cycle*, Boca Raton, FL: CRC Press, pp. 459–71.

Paustian, K., H.P. Collins and E.A. Paul (1997b), 'Management controls on soil carbon', in E.A. Paul, K. Paustian, E.T. Elliott and C.V. Cole (eds), *Soil Organic Matter in Temperate Agroecosystems: Long-term Experiments in North America*, Boca Raton, FL: CRC Press, pp. 15–49.

Paustian, K., E.T. Elliott, J. Six and H.W. Hunt (2000), 'Management options for reducing CO_2 emissions from agricultural soils', *Biogeochemistry*, **48**: 147–63.

Pautsch, G.R., L.A. Kurkalova, B.A. Babcock and C.L. Kling (2001), 'The efficiency of sequestering carbon in agricultural soils', *Contemporary Economic Policy*, **19**(2): 123–34.

Peterson, G.A., A.D. Halvorson, J.L. Havlin, O.R. Jones, D.J. Lyon and D.L. Tanaka (1998), 'Reduced tillage and increasing cropping intensity in the Great Plains conserves soil C', *Soil Tillage Research*, **47**: 207–18.

Plantinga, A.J., T. Mauldin and D.J. Miller (1999), 'An econometric analysis of the cost of sequestering carbon in forests', *American Journal of Agricultural Economics*, **81**(4): 812–24.

Rosenzweig, C., B. Curry, J.T. Ritchie, J.W. Jones, T.Y. Chou, R. Goldberg and A. Iglesias (1994), 'The effects of potential climate change on simulated grain crops in the United States', in C. Rosenzweig and A. Iglesias (eds), *Implications of Climate Change for International Agriculture: Crop Modeling Study* (EPA 230-B-94-003), Washington, DC: US Environmental Protection Agency.

Schimmelpfennig, D. and G. Yohe (1999), 'Vulnerability of crops to climate change: a practical method of indexing', in G. Frisvold and B. Kuhn (eds),

Global Environmental Change and Agriculture: Assessing the Impacts,
Cheltenham, UK: Edward Elgar Publishing, pp. 193–217.

Schneider, S.H. (1997), 'Integrated assessment modeling of global climate
change: Transparent rational tool for policy making or opaque screen
hiding value-laden assumptions?', *Environmental Modeling &
Assessment*, **2**(4): 229–48.

Soil Conservation Service (SCS) (1981), *Land Resource Regions and Major
Land Resource Areas of the United States, USDA Agriculture Handbook
296*, Lincoln, NE: United States Department of Agriculture, Soil
Conservation Service, National Soil Survey Center, pp. 156.

Soil Conservation Service (SCS) (1994), *State Soil Geographic Data Base
(STATSGO) Data User Guide*, Lincoln, NE: United States Department of
Agriculture, Soil Conservation Service, National Soil Survey Center.

Segerson, K. and B.L. Dixon (1998), 'Climate change and agriculture: the role
of farmer adaptation', in R. Mendelsohn and J.E. Neumann (eds), *The
Impact of Climate Change on the US Economy*, Cambridge: Cambridge
University Press, Chapter 4.

Stavins, R.N. (1999), 'The costs of carbon sequestration: a revealed-
preference approach', *American Economic Review*, **89**(4): 994–1009.

VEMAP (1995), 'Vegetation/ecosystem modeling and analysis project
(VEMAP): comparing biogeography and biogeochemistry models in a
continental-scale study of terrestrial ecosystem responses to climate
change and CO_2 doubling', *Global Biogeochemical Cycles*, **9**: 407–37.

6 Assessing Potential Public Health Impacts of Changing Climate and Land Uses: The New York Climate and Health Project

**P. Kinney, J. Rosenthal, C. Rosenzweig,
C. Hogrefe, W. Solecki, K. Knowlton,
C. Small, B. Lynn, K. Civerolo, J. Ku,
R. Goldberg and C. Oliveri**

INTRODUCTION

Over the next 50 years, a rapidly urbanizing world population will confront significant environmental change caused by a warming climate and rapid conversion of land. The Intergovernmental Panel on Climate Change (IPCC) Third Assessment projects that the globally averaged surface temperature will increase by 1.4–5.8°C (2.4–10.4°F) by 2100 (IPCC 2001).[1] Simultaneously, human populations are carrying out a rapid and substantial conversion of land from natural to human dominated uses. To be responsible stewards of both human health and biological diversity in the coming century, societies will need to develop and institutionalize better models describing and predicting the interactions between these global drivers and the health of the planet. The objective of the project described here is to begin to build and apply a modeling framework that assesses potential future public health impacts of both climate change and land use change in the New York metropolitan region.

There are many challenges to overcome in addressing this objective. One is the need to bring together and integrate a diverse set of models and observations, including those addressing changes in global climate (General Circulation Models or GCMs), land use, regional climate, air quality and human health. Another important technical challenge is the need to take the broad scale predictions generated by GCMs, and make them relevant and meaningful for impact assessments at finer geographic scales. Outputs from GCMs typically are resolved at a scale of hundreds of kilometers. Development and integration of research on the human dimensions of global environmental change require new methods for downscaling coarse-resolution global climate, atmospheric chemistry models and demographic projections to the regional metropolitan scale (tens of miles/kilometers or finer). In order to predict and plan for the impacts of climate change, there is a need to assess the possible impacts of changes in temperature and air quality at the regional and urban scales, where the planning for mitigation and adaptation strategies may take place.

With funding from the Environmental Protection Agency (EPA) STAR program, the project 'Modeling Heat and Air Quality Impacts of Changing Urban Land Uses and Climate' is designed to address these challenges.[2] Hereafter referred to as the New York Climate and Health Project (NYCHP), the project has developed an integrated modeling framework capable of providing regional projections of climate and air quality under alternative scenarios of global climate change and regional land use change. The integrated framework is being used to assess potential public health impacts of both extreme heat events and ozone air quality over the coming century in the New York metropolitan region.

For scenario-based, integrated health risk assessments there are several sources of uncertainty in estimating future impacts. The climate and air quality models introduce uncertainty into the system, yet can test their simulations against meteorological data to find the degree to which the models capture the observations. The underlying population mortality rates alter in response to many demographic, social, behavioral and political factors regarding individual and group health and access to health care. The climate-human health relationship within a given geography and population may change over time if populations acclimate and/or adapt to changing conditions. The interdisciplinary process involved in downscaling from global to local impacts, involves simplifications in a particular team's modeling methods that will introduce uncertainties. From a health science perspective, using a variety of concentration-response functions and their associated confidence intervals from the relevant epidemiological literature is one method for expressing uncertainty. By anticipating the range of possible impacts, the range of possibilities suggested by each scenario's environmental, technological, demographic, socioeconomic and political storyline can be examined.

This chapter outlines the objectives, design and methods used in the integrated modeling framework developed by the NYCHP and illustrates the kinds of public health impact assessments that are being generated. We also discuss potential strategies for mitigating these impacts.

HEALTH CHARACTERISTICS OF THE NEW YORK METROPOLITAN REGION

The New York Metropolitan Region (NYMR) has been defined by the Regional Plan Association as the 31-county, three-state region shown in Figure 6.1 (RPA 1996). With New York City (NYC) at its core, this 33 600 square kilometer (13 000 square mile) region is presently home to over 21 million people. It has a widely varying landscape and a range of population densities and land uses. In addition to the nation's largest city, the metropolitan region includes a relatively pristine watershed, source of NYC's drinking water; substantial agricultural land in parts of northern New York, Long Island and central New Jersey; and an estimated 1600 cities, towns and villages. The cultural, ethnic, racial and socioeconomic diversity of the region makes it unique among world metropolitan areas. This is mirrored in a complex set of public health vulnerabilities. These include extensive areas of extreme poverty, particularly in the inner city; high population density; the constant in-flux of immigrants and transients; an unfiltered municipal water supply; and a sizeable population of immuno-compromised persons (Hamburg 1998).

Heart disease is the leading cause of death in both the NYMR and the US as a whole. The mortality rate for HIV is over twice as high in the NYMR as in the rest of the country. Asthma mortality is similar in both the NYMR and the US as a whole; however, large variations exist across counties within the region.

Asthma hospitalization rates have increased nationwide over the past three decades, particularly in the northeast (Mannino et al. 1998). In boys and girls under age 15, asthma has ranked second and third, respectively, for causes of hospitalization among Connecticut residents (CT DPH 1998) and ranks first in New York City, where the childhood rates are three times the national average (NYC DOH 1998). Hospitalization rates for asthma and total respiratory causes (including asthma, bronchitis, pneumonia and emphysema) for selected NYMR counties in 1996 are displayed in Table 6.1. These data demonstrate widely varying rates across counties similar to those observed for asthma mortality. In 1996 Bronx County, NY had over four times the national asthma hospitalization rate; parts of Manhattan (for example East Harlem) led the nation with seven times the national rate. Within the city, asthma hospitalization rates for children in poor minority neighborhoods are almost three times greater than for children in high-income neighborhoods, while four times as many adults went to the emergency room in poorer than in high-income neighborhoods (NYC DOH 2003).

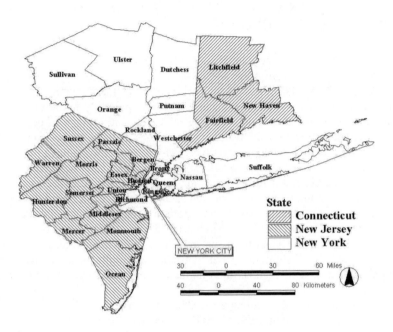

Source: RPA 1996

Figure 6.1   *New York Metropolitan Region as defined by the Regional Plan Association*

CURRENT PUBLIC HEALTH STRESSORS

Chief among the current public health vulnerabilities in the New York Metropolitan Region is poverty, which is endemic to pockets of many counties. The spatial distribution of income within NYC is displayed in Figure 6.2. While dilapidated housing and poor indoor air quality contribute to the adverse health characteristics of disadvantaged neighborhoods (Gold 1992; NAS 2000), outdoor air quality also plays a role.

While considerable progress has been made over the past 30 years in reducing levels of some outdoor air pollutants (for example sulfur dioxide and carbon monoxide) in the NYMR, other pollutants, especially ozone and particulate matter, continue to reach unhealthy levels on a regular basis (US EPA 2003a, 2003b, 2004). The human-health based National Ambient Air Quality Standards (NAAQS) for ozone and particulate matter are often exceeded in the region, placing several counties out of compliance. Human health effects that have been associated with these two pollutants include mortality and hospitalizations for cardiovascular and respiratory diseases, increases in respiratory symptoms such as cough and wheeze, and diminished lung function (Kinney 1999). Effects are greatest among the elderly, the young and persons with compromised health status such as asthmatics.

Another outdoor pollutant of concern, but for which no standard exists, is diesel exhaust particles (DEP), emitted in large quantities by trucks and buses throughout much of the NYMR. Diesel particles consist of tiny carbonaceous nuclei upon which are adsorbed a wide variety of organic compounds, including the carcinogenic polycyclic aromatic hydrocarbons (Kinney et al. 2000). Because of their small size, DEP can be inhaled and deposited deeply in the human respiratory tract. Occupational epidemiology studies have linked DEP exposures with lung cancer. Other studies have linked exposure to

Table 6.1   *Hospitalization rates for total respiratory conditions and asthma, selected NYMR counties and the US, 1996 (hospitalizations/100 000 persons/year)*

	Total Respiratory	Asthma
Connecticut		
Fairfield	748.3	130.9
Litchfield	874.5	85.6
New Haven	982.9	167.5
New York		
Bronx	1964.7	846.9
Dutchess	873.2	105.1
Kings	1544.5	511.8
Nassau	877.6	160.7
New York	1256.8	420.7
Orange	1133.6	198.3
Putnam		
Queens	1083.4	281.9
Richmond	1270.4	246.6
Rockland	644.3	98.8
Suffolk	769.1	144.7
Sullivan	1194.2	133.6
Ulster	985.9	104.4
Westchester	814.8	124.1
US	1226.5	179.5

Source: INFOSHARE geographic information systems, developed by Community Studies, Inc.

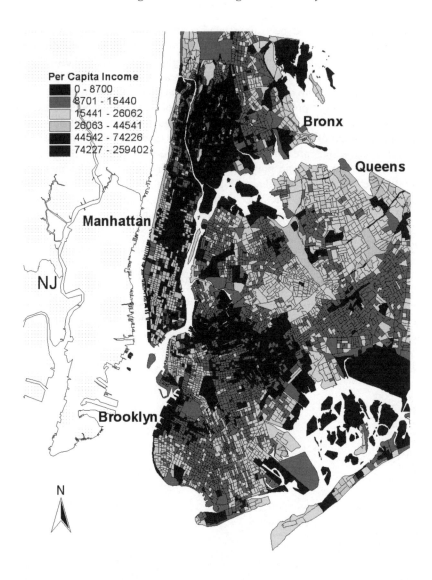

Figure 6.2   Geographic distribution of income per capita across New York
City in 1990 (data aggregated at the census block group level)

traffic-related pollution (such as that resulting from residential proximity to major roadways) with increased respiratory symptom rates and diminished lung function (see review in Kinney et al. 2000). Within the NYMR, DEP exposure is often viewed as an environmental justice issue with respect to the locations of bus depots and other diesel-related sources. For example, six of seven bus depots and stations in Manhattan are located north of 100th Street in the underprivileged and largely minority communities of Harlem and Washington Heights (Northridge et al. 1999).

In New York, as in other cities around the world, summertime heat can lead to elevated mortality and morbidity rates, especially during the extended periods of hot weather known as heatwaves. Since 1998, summertime heat has been the top weather-related cause of mortality in the United States (NOAA 2003). Numerous epidemiology studies have examined the relationship between extreme heat events and increases in short-term mortality in urban populations in the temperate zone (for example Kalkstein and Greene 1997; Braga et al. 2002).

The epidemiological literature has identified factors in the built environment and demographic characteristics that can increase the risk of heat-related mortality. The elderly and people with pre-existing illnesses are especially vulnerable populations as are those persons who are bedridden, live alone and/or have poor access to public transportation or air conditioned neighborhood places. Analysis of the Chicago 1995 heatwave, which led to over 700 excess deaths, showed that risk of heat-related mortality was higher in the black community, and in those living in certain types of low-income and multi-tenant housing, including living on the top floor of buildings (Klinenberg 2002).

RESEARCH FRAMEWORK FOR THE NYCHP: OBJECTIVES, APPROACH AND PROCESS

Objectives

Heatwaves and elevated concentrations of ozone are two significant current public health stressors in the New York Metropolitan Region. Both of these stressors may be exacerbated by future changes in the global climate as well as continued expansion of human-dominated land uses in the region. As we have seen, numerous existing environmental conditions and health stressors exist in the NYMR, and these may lead to highly variable vulnerabilities to new stressors. Tools are needed that make it possible to assess potential future health impacts of heat and air quality under alternative scenarios of climate and land use changes, and which can be applied at the fine geographic scales relevant to regional decision making.

The objective of the NYCHP is to develop an integrated modeling framework for assessing public health impacts of changing climate and land uses at a regional scale in the New York Metropolitan Region. To accomplish this objective, we linked the Goddard Institute for Space Studies (GISS) Global Climate Model (GCM), the Penn State/National Center for Atmospheric Research (PSU/NCAR) MM5 meso-scale meteorological models, the SLEUTH land use/land change (LU/LC) model, the Community Multiscale Air Quality model (CMAQ) and a public health risk assessment. The linkages among these models are displayed in Figure 6.3. The project is guided by four questions:

1. What changes in the frequency and severity of extreme heat events are likely to occur over the next 80 years due to a range of possible scenarios of land use/land cover and climate change in the New York Metropolitan Region?
2. How might the frequency and severity of episodic concentrations of ozone change over the next 80 years due to a range of possible scenarios of land use and climate change in the region?
3. What is the range of possible human health impacts of these changes?
4. How might projected future human exposures and responses to heat stress and air quality differ as a function of socioeconomic status and race/ethnicity across the region?

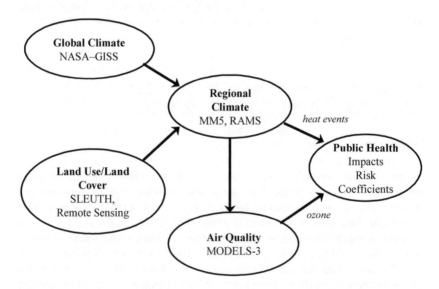

Figure 6.3   The integrated modeling framework of the New York Climate and Health Project

Approach

To structure our work we established a set of protocols. First, future scenarios of greenhouse gas emissions, ozone precursor emissions and land use changes were defined on the basis of IPCC Special Report on Emissions Scenarios (SRES) A2 and B2 scenarios. The SRES scenarios represent a set of storylines for possible future development (Table 6.2) (based on Nakićenović et al. 2000). The A2 scenario is characterized by high CO_2 emissions (30 gt/yr max), relatively weak environmental concerns and large population increases (15 billion by 2100). Economic development is primarily regionally-oriented and per capita economic growth and technological change more fragmented and slower than other storylines. The B2 scenario is characterized by medium CO_2 emissions (15gt/yr max), strong emphasis on environmental issues and medium population growth (10 billion by 2100). It is a world with continuously increasing global population, at a rate lower than A2, intermediate levels of economic development, and less rapid and more diverse technological change than in the B1 and A1 storylines.

Second, for modeling runs downscaled from the global to the regional and urban scales, we defined a modeling domain that includes the eastern half of the US (Figure 6.4). This domain was used for climate and air quality simulations at 36×36 km grid size, and made it possible to account for the upwind meteorological and air pollution source regions relevant to impacts in the NYMR. As noted earlier, the health impacts domain was defined as the 31 county NYMR (Figures 6.1 and 6.4). Within the NYMR domain, we assessed impacts at a range of grid scales, with the finest scale being 4x4 km.

A third set of key protocols relates to the time slices to be modeled. GCM outputs were available on an hourly basis from the 1990s through the 2080s. Regional climate and air quality simulations on 36 km grids over the eastern US domain were carried out for the summer months of June–August for five consecutive years mid-decade (for example 1993–97) of the 1990s, 2020s, 2050s and 2080s. For finer resolution simulations at 12 and 4 km within the NYMR, episodes of high temperature and ozone levels were selected. For assessing public health impacts of heat events, we focused the analysis on the effects related to premature mortality. For public health impacts of ozone, we focused on premature mortality and on excess hospitalizations for respiratory diseases.

The five research areas identified on Figure 6.3 represent active research communities that typically work independently of one another. The NYCHP relies on collaborative interaction among these research teams to create climate projections specifically tailored to physical conditions and human populations in the 31-county NYMR study area. Furthermore, all five modeling teams are unified in their use of the A2 (rapid growth) and B2 (slower growth) scenarios.

Table 6.2   *Summary of key features of A2 and B2 scenarios of future growth*

Parameter	Scenario	
	A2 HIGH-CO_2 (30 gt/yr max)	B2 MEDIUM-CO_2 (15 gt/yr max)
World	Differentiated, divided world; economy drives	Increased concern for environmental and social sustainability; regional stewardship; econ/ecol values
Capital	Lower trade flows; slower capital stock turnover	Relatively slow rate of development especially in developing nations
Technology	Slower technological change; heterogeneity; fast in some places, slow in others	Technological change still uneven; mechanisms for international diffusion of technology and knowledge higher than in A2, yet weaker than A1 or B1
Economy	'Consolidates' into series of economic regions; economic growth uneven; income gap does not narrow between now-industrial versus developing nations; world GDP US$250 trillion by 2100	Stronger community support networks; local inequity is reduced considerably; international income differences decrease; world GDP US$250 trillion by 2100
Population	Largest population increase: 15 billion by 2100; emphasis on family and community life; fertility rates decline slowly	Moderate population growth: 10 billion by 2100; (UN 1998 base 'medium projection')
Mobility and land use	Less mobile (people, capital, ideas); urban sprawl	Urban and transportation infrastructure is a particular focus of community innovation; low level of car dependence; less urban sprawl; land use becomes better integrated at local level
Per capita income	Lower per capita income growth: average world PCI low: US$7200 by 2050; US$16K by 2100	'Moderate' economic growth; more affluent; average world PCI intermediate: US$12K by 2050

(continued on next page)

Table 6.2 continued

| Parameter | Scenario | |
| | A2 | B2 |
	HIGH-CO2 (30 gt/yr max)	MEDIUM-CO2 (15 gt/yr max)
Energy	Resource availability determines fuel mix; fossil fuel use relatively greater; energy intensity declines slowly, by 0.5–0.7%/y	Hydrocarbon-based energy to 2100, yet gradual transition away from fossil fuels; energy intensity of GDP declines 1%/yr (in line with average experience since 1800)
Food	Agricultural productivity is a main focus; locally sustainable, high yields after initial period of soil erosion and water pollution	Emphasis on local food reliance; lower meat consumption in nations with high population densities
Environment	Environmental concerns relatively weak; regional and local control structure; average summertime ozone levels increase by 30 ppb in northern mid-latitudes	Environmental and human welfare are high priorities (especially at regional/local level); less carbon-intensive technologies, better regional management decreases emissions of SO_2, NO_x, VOCs and ozone
Governments	Social and political structures diversify; protectionism among nations; diversity, yet conflicts between civilizations rather than globalizing economies may determine the geopolitical future of the world	Decentralized governments and international institutions decline in importance except for environmental protection

Process

Global climate modeling

Global climate modeling was carried out using the GISS coupled atmospheric-ocean model, version III, with a grid resolution of 4° × 5° latitude and longitude. In this model computations are made for nine vertical atmospheric layers and 13 vertical ocean layers with realistic bathymetry (Russell et al. 2003). Global climate was simulated for the period 1850–2100 with the IPCC A2 and B2 greenhouse gas (GHG) forcings. These include projected changes in CO_2, CH_4, N_2O, sulfates, CFC11 and CFC12.

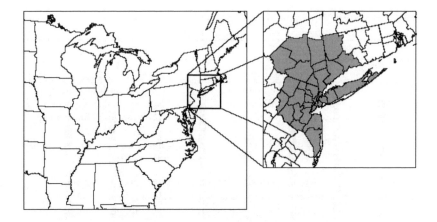

Note: The 31-county New York Metropolitan Region is highlighted in the insert.

Figure 6.4   Map of the 36 km CMAQ modeling domain

Full sets of GCM variables from these simulations for the selected 5-year time-periods were provided to the regional climate model (RCM) groups for use as boundary conditions in their simulations. In addition, the GCM outputs for daily maximum temperature at the Earth's surface were used in health impact analyses within the NYMR.

Regional climate modeling
The MM5 regional climate model was used in the project to simulate the current and future meteorology of the New York metropolitan region. The PSU/NCAR meso-scale model (known as MM5) is a limited-area, non-hydrostatic, terrain-following sigma-coordinate model designed to simulate or predict meso-scale and regional-scale atmospheric circulation. It was developed at Penn State and NCAR as a community meso-scale model and is continuously being improved by contributions from users at universities and government laboratories. The Fifth-Generation NCAR/Penn State Mesoscale Model (MM5) is the latest in a series that developed from a meso-scale model used by Anthes at Penn State in the early 1970s that was later documented by Anthes and Warner (1978). Since that time, it has undergone many changes designed to broaden its usage. These changes include the addition of:

1. a multiple-nest capability;
2. non-hydrostatic dynamics, which allows the model to be used at a few-kilometer scale;
3. multitasking capability on shared- and distributed-memory machines;

4.   a four-dimensional data-assimilation capability; and
5.   more physics options.

In the present project, we utilize MM5 simulations driven by the GISS-GCM through boundary and initial condition inputs that were performed for the summer seasons (June–August) for five consecutive mid-decadal years (for example 1993–1997) in the 1990s, 2020s, 2050s and 2080s. The MM5 was applied in a nested-grid mode with an inner-grid having a horizontal resolution of 36 km over the eastern US and an outer-grid having a horizontal resolution of 108 km covering most of the continental US. The MM5 had 35 vertical layers; the height of the first layer was approximately 70 meters. Lynn et al. (2004) tested several different combinations of MM5 physics options. For this study we selected the MM5 simulations that were performed with the medium range forecast (MRF) boundary layer scheme (Hong and Pan 1996), the Betts–Miller cloud scheme (Betts 1986) and the rapid radiative transfer model (RRTM) radiation scheme (Mlawer et al. 1997). Extensive evaluation and validation was performed at the 108 km scale (Lynn et al. 2004). Figure 6.5 shows observed and simulated surface temperatures for the 1990s, generated from the GISS-GCM and by MM5 running at 108 and 36 km scales. The 36 km simulations capture some of the spatial detail that is missed at lower resolution, such as over the Appalachian Mountains. Results from the 36 km simulation were used to evaluate regional health impacts of temperature extremes and were used to perform air quality simulations.

Figure 6.6 plots surface temperature projections over Central Park for three decades of the 21st century under the A2 and B2 SRES scenarios expressed as changes from model estimates for the 1990s. Decadal values represent tri-decadal averages (for example 2020s = average of the 2010s, 2020s and the 2030s). Results show progressive regional warming in both the A2 and B2 scenarios, with stronger climate effects present in the A2 scenario with the greater GHG forcing. By the 2080s, the projected temperature increases for the region range from 2.0–2.5°C in B2 scenario and 3.0–3.5°C in the A2 scenario.

Land use modeling
We adapted a computer-based land use change model to represent future scenarios of urban change in the New York metropolitan region. The land use model utilized in the project is the SLEUTH program (Clarke et al. 1997). The SRES A2 and B2 scenarios were used to define the potential conditions of future land use change. The land use change scenario experiments developed for the study represent an innovative application of the SLEUTH program. The case study region of the project for land use modeling is the 31-county NYMR.

Simulated Summer 1990s Temperature

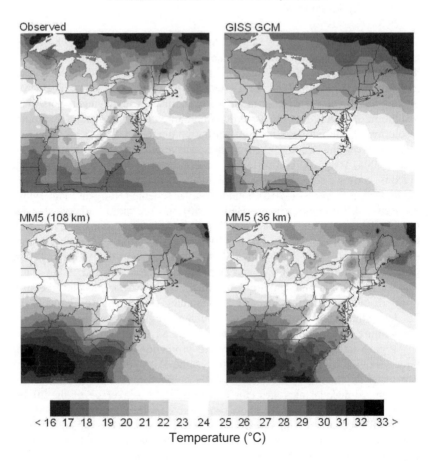

Source: Observed and simulated surface temperatures for the 1990s at alternative scales

Figure 6.5   Observed and simulated surface temperatures for the 1990s at alternative scales

As global change science emerged in the past decade there has been a concurrent growing appreciation of the importance of the land use/land cover measures in the understanding global environmental change and the generation of GHG. The publication of the IPCC SRES narratives

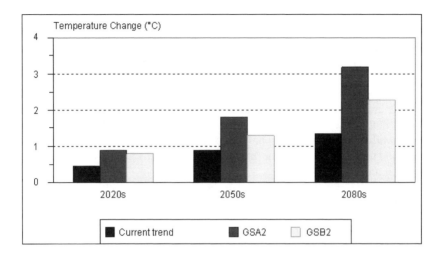

Figure 6.6   *Projected change in annual climate in three decades of the 21st century compared to the 1990s, showing extrapolation of the current trend and the outputs of the GISS-GCM forced with SRES A2 and B2 emissions*

(Nakićenović et. al. 2000) and similar reports (for example US EPA 2003c) have helped increase the focus on the role of urban settlements and associated urban land use in the GHG emissions. Different urban land use patterns and population densities have been used as measures and indices of different GHG profiles (Gaffin 1998, 2002). Low density, urban sprawl communities are often defined as relatively high-per capita GHG emissions sites (Liu et. al. 2003). A key component of the Climate and Health Project is to illustrate the role and effectiveness of computer-based urban land use modeling activities in broader scale interdisciplinary climate change studies.

This work had two main objectives. First we aimed to define the projected conversion probabilities and the amount of rural-to-urban land use change for the NYMR as derived by the Urban Growth Model (UGM) and Land Cover Deltatron Model (LCDM) for the years 2020 and 2050, as defined by the pattern of growth for the years 1960 to 1990. Second, we sought to down-scale the IPCC SRES A2 and B2 scenarios as a narrative that could be translated into alternative growth projections. The modeling would be done for the years 2020 and 2050.

SLEUTH is a probabilistic cellular automata model that defines future land use change as a product of a set of growth inducing variables (such as slope, land cover, exclusion zones, land use, transportation and hillshading), a set of growth parameters (defined by the past patterns of urbanization) and growth rules. The software structure enables one to define future growth as a projection of past growth, as well as define alternative growth scenarios (such as slowed conversion or more rapid conversion).

SLEUTH is comprised of the UGM and the LCDM. The UGM simulates land class change, or more specifically, the probability that a non-urban cell will be converted to an urban cell. The LCDM, which is driven by the UGM, include land cover data in the simulation and therefore can specify the nature of the non-urban to urban changes (for example the amount of agricultural land to urban land change, the amount of forest land to urban land change and so on). The UGM may be run independently of the LCDM, but the LCDM must be run with the UGM.

In order to create the range of possible climate outcomes, a number of land use change scenarios utilizing the SLEUTH program were developed. The analysis first examined past changes in urban land cover and then applied these trends to construct a range of scenarios for future land use changes in the NYMR for the years 2020 and 2050. The researchers used both the SLEUTH UGM and LCDM programs and a set of land use and land cover data from the 1960s to 1990 to create the scenarios of land use/cover change. Before future land use scenarios were generated, model results for the 1960 to 1990 period were compared against observed data of land use change for the same period in order to validate the model structure.

Preliminary modeling work suggests that there could be significant land use and land cover change in the NYMR during the first half of the 21st century. For example, the UGM model projects a loss of 47 percent and 67 percent in 2020 and 2050, respectively, of the total non-urban land present in 1990. The rate of conversation slows significantly during the 2020 to 2050 period because the number of available sites (pixels) for spontaneous and new breeding centers becomes limited and instead an increased proportion of the new growth takes place as slower edge growth or transportation corridor related growth. Projected rapid conversion takes place where significant conversion had occurred during the period 1960 to 1990 and where the development potential was high (for example areas with relatively flat terrain or access to highways). As a result, conversion was particularly extensive in eastern Long Island and central New Jersey. The more mountainous and isolated northern parts of the region incurred less development during the study period.

The model results have utility for public health exposure and risk analysis. Project work is being planned to use the urban land use model results to develop estimates of future population growth in the sub-urbanizing parts of the region at the US census tract level (spatial areas defined by approximately 5000 residents). The amount of urban land development in each tract will be associated to a relative increase in population. Another use to which the land use projections are being put is to alter land surface parameters used as inputs to the regional climate and air quality models.

Regional ozone modeling

For air quality simulations, we use the Community Multiscale Air Quality (CMAQ) model (Byun and Ching 1999) with the Sparse Matrix Operator Kernel Emissions Modeling System (SMOKE) (Carolina Environmental Programs 2003). The GCM/MM5 linked model provides the meteorological inputs needed for the air quality simulations. The outputs from the air quality simulations have been used to validate the modeling system against observed ozone data (Hogrefe et al. 2004a), to project future ozone throughout the 36 km eastern US modeling domain (Hogrefe et al. 2004b) and for assessing potential public health impacts based within the NYMR (Knowlton et al. 2004).

Climate change can influence the concentration and distribution of air pollutants through a variety of direct and indirect processes, including the modification of biogenic emissions, the change of chemical reaction rates, mixed-layer heights that affect vertical mixing of pollutants, and modifications of synoptic flow patterns that govern pollutant transport. For example, warmer temperatures can result in increase of concentrations of photochemical oxidants (Sillman and Sampson 1995), while many past studies have revealed the impact of meteorological conditions on episodes of high ozone concentrations (Gaza 1998; Seaman and Michelson 2000) which, in turn, can aggravate respiratory disorders (Schwartz and Dockery 1992; Katsouyanni et al. 1993).

The county-level EPA 1996 National Emissions Trends (NET96) inventory was used for anthropogenic emission estimates. This emission inventory was processed by the SMOKE modeling system (Carolina Environmental Programs 2003) to obtain gridded, hourly, speciated emission inputs for the air quality model. Future year emissions are estimated by multiplying the base year emissions with the regional growth factors for the SRES A2 scenario. Temperature-dependent biogenic emissions were estimated by the Biogenic Emissions Inventory System – Version 2 (BEIS2) (Geron et al. 1994; Williams et al. 1992), while mobile source emissions were estimated by the Mobile5b model (US EPA 1994).

Using this modeling system under the A2 SRES scenario, we estimated hourly surface ozone concentrations over the 36 km domain for five mid-decadal summers in the 1990s and 2050s (Figure 6.7). This figure illustrates changes that may occur in ozone concentrations due purely to projected climate changes (top right panel), due purely to projected emissions changes (bottom left panel) and due to the combined effects of changes in both drivers (bottom right panel).

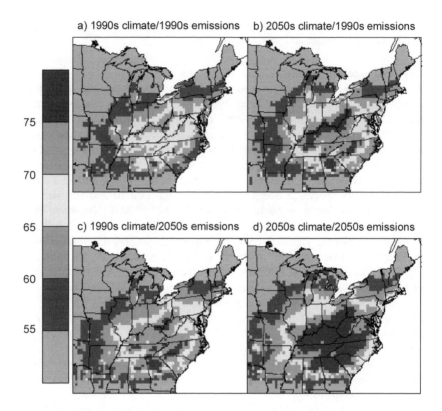

*Figure 6.7   Summertime daily maximum 1-hour ozone concentrations,
                 averaged over five mid-decadal years, from simulations with
                 current/future climate/emissions*

The GISS-GCM-MM5-CMAQ modeling system was evaluated with observations for the five summers of 1993–97, in order to understand how well it simulated present climate and current ozone concentrations on a regional-scale over the eastern United States and over the metropolitan region.

The MM5 simulations, driven by the GISS-GCM through boundary and initial conditions inputs performed for the five 1990s summers, was applied in a nested-grid mode with an inner-grid having a horizontal resolution of 36 km over the eastern US. Results from the 36 km simulations were used to perform CMAQ simulations. Results indicate that MM5 and CMAQ capture the inter-annual and synoptic-scale variability present in observations in the current climate, while the magnitude of fluctuations on the intra-day and diurnal time scales is underestimated (Hogrefe et al. 2004a). The spatial pattern of five summer average daily maximum 1-hour ozone concentrations is reproduced by CMAQ. The frequency distribution of the duration of extreme heat and ozone events is also generally well captured by MM5 and CMAQ, and the modeling system succeeded in simulating the mean ozone concentrations associated with several frequent pressure patterns well, indicating that both the mean emission patterns and the effects of synoptic-scale meteorology on ozone concentrations are well-represented in the model ensemble.

PUBLIC HEALTH IMPACT ASSESSMENT

Having identified heat stress and ozone air quality as two current health impacts of concern, the driving research objective of the NYCHP is to project associated health impacts of climate change for the New York metropolitan region. The project attempts to respond to a need for regional projections of these possible future health impacts over the medium (25 years) to long term (100 years). In the past, the potential effects of climate change have been projected in terms of global or continental-scale temperature and precipitation variation at scales of hundreds of kilometers. To date however, specific public health impacts resulting from environmental change have seldom been projected at the regional level more appropriate to adaptive city, county and state planning. Public health impact assessments under the current project are being performed for both temperature and ozone. Future extensions may incorporate other parameters of weather and additional pollutants such as fine particulate matter.

Public health impacts are estimated using a risk assessment framework. To illustrate, the formula below was used to estimate increases in daily mortality for a specific population (city, county or region) in a given time period (1990s versus future decade). Note that daily impacts were summed within the time period of interest.

$$A = (P/100\ 000) * M * CR * E$$

A: estimated number of additional daily deaths attributable to environmental change (temperature or ozone)

P: estimated resident population during time period of interest

M: baseline daily mortality in the absence of environmental change (per 100 000 population)

CR: concentration response; a coefficient which quantifies the magnitude of the proportional change in mortality that would be expected in response to a unit change in environmental conditions (CR coefficients are regression slopes reported in the existing epidemiologic literature)

E: daily concentration of the environmental variable. When a threshold concentration is assumed, this is the amount by which the daily concentration exceeds the threshold concentration.

The 2000 US Census totals for each of the 31 counties in the study area were used as the 'reference year' or baseline populations, which were then aggregated to approximate the current regional population (approximately 21.5 million). County and aggregate regional population growth during the 2020s, 2050s and 2080s were projected by downscaling A2-consistent US growth projections through the year 2100 (Gaffin 2002). This method projected a 24 percent increase in regional population by 2025, a 53 percent increase by 2055 and a 95 percent increase by 2085. Applying the A2 regional growth figures within the health risk assessment provided a means of looking at the effects of population growth on mortality and morbidity according to a more reasonable A2 'alternate future'.

Typical 1990s daily summer mortality rates for each of the 31 counties in the study area were estimated using all-age crude mortality data for all internal causes (ICD-9 codes 0–799.9 for years 1990–98, and ICD-10 codes A00-R99 for year 1999) from the US Centers for Disease Control and Prevention (US CDC 2004).

Health impacts were assessed for ozone air quality (from simulated 1-hour daily maximum ozone concentrations, in ppb) and ambient temperatures (as daily mean temperatures in °F). Future changes in these two environmental stressors as simulated by the NYCHP model system were compared relative to conditions in the 1990s across the NY metro region, as taken from meteorological observations at the GCM scale or from interpolated CMAQ ozone and/or MM5 temperature simulations at the RCM scale.

CR coefficients were taken from the recent environmental epidemiology literature. For the temperature-mortality relationship, the CR was taken from the NYC results from a study of 11 eastern US cities (Curriero et al. 2002, 2003). For ozone and mortality, we drew our CR from a meta-analysis of past observational epidemiological studies on acute ozone mortality, which found a pooled effect estimate of RR 1.056 (95 percent confidence interval 1.032–1.081) associated with a 100 ppb increase in daily 1-hour maximum ozone (Thurston and Ito 1999, 2001). Seven major impact analyses emerged from the project:

1. ozone-related mortality and morbidity;
2. heat-related mortality and morbidity;
3. relative summer versus winter mortality changes;
4. differences in health impacts projected using different spatial scales (global versus regional) in future climate model projections;

5. differences in health impacts projected using different temporal scales (multi-day heatwaves versus individual hot days) in future climate model projections;
6. examination of the distribution of impacts among vulnerable populations in selected case study areas; and
7. the effect of A2- and B2-consistent regional land use changes upon projected temperature and ozone profiles and corresponding health impacts.

The objective of the overall study was to compare the relative effects of each type of variation upon projected regional mortality and morbidity in a 'meta-sensitivity analysis'.

To illustrate preliminary health impact results for ozone, projected increases in annual county-specific ozone-related deaths for the 2050s versus 1990s under A2 are shown in Figure 6.8. The counties downwind and to the east (in this case Long Island and Connecticut, less densely populated counties) of ozone precursor source areas are projected to experience relatively greater increases in summer ozone mortality than the urban core counties of New York City. This shows that the pattern of relatively high summer ozone areas in future years does not fully coincide with population centers, thereby diluting ozone's public health impact. Although percent increases in ozone-related mortality are projected to be greatest in the counties up-wind and down-wind of NYC, absolute numbers of ozone-related deaths would be greatest in the urban counties where population is highest.

Regional heat-related summer mortality projections in the 2050s under A2 as compared to the 1990s are illustrated in Figure 6.9. This figure shows that in the 36 km grid resolution MM5 simulations, the pattern of relatively high summer heat areas in future years coincides with population centers, thereby amplifying heat's public health impact. Also note that the temperature-related impacts appear larger than those estimated for ozone above.

To evaluate the effects of analysis scale on projected health impacts, we compared impacts of temperature on daily deaths derived using the 36 km GISS/MM5 modeling framework to those derived at the 4 x 5° scale using just the GISS-GCM results for the NYMR. Though both models projected increased heat-related deaths in the 2050s, the 36 km analysis resulted in larger impacts. This was likely due to the positive interaction between temperature increases and population densities across the region. These results serve to illustrate one important way in which scale can matter in impact assessment.

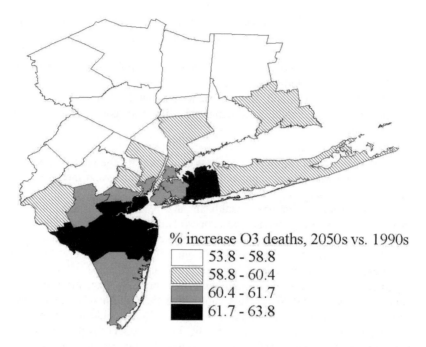

% increase O3 deaths, 2050s vs. 1990s
- 53.8 - 58.8
- 58.8 - 60.4
- 60.4 - 61.7
- 61.7 - 63.8

Notes: No changes in anthropogenic ozone precursor emissions are assumed here for future decades, and therefore all increases in ambient ozone are attributable to climate change.

Figure 6.8   Estimated percent increases in ozone-related deaths for a typical summer of the 2050s under the A2 scenario versus ozone-related mortality for a typical summer of the 1990s

FUTURE NYCHP HEALTH IMPACTS ANALYSES

The NYCHP analysis concentrated on the 2050s as a target decade during which to evaluate ozone- and temperature-related mortality. The IPCC SRES A2 scenario was chosen because it represents one of the more extreme growth scenarios, thus representing a possible extent of the full 'envelope' of impacts to consider.

Future public health impacts analyses will include evaluating morbidity changes related to heat and ozone (for example all-cause and asthma hospitalization rate changes); evaluating multi-day high temperature episodes (such as heatwave impacts); examining the distribution of impacts among

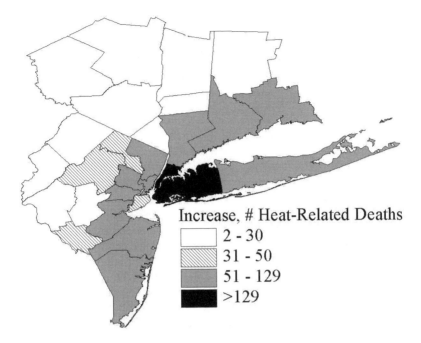

Figure 6.9   Map of increases in numbers of heat-related deaths projected for a typical summer of the 2050s under the A2 scenario versus heat-related deaths typical of the 1990s

vulnerable populations (the elderly, the urban poor and so on) across the region; using the B2 (slower-growth) IPCC scenario throughout the model system; downscaling further to 12 km and 4 km grid resolution in the MM5 and CMAQ outputs; and evaluating the effect of A2- and B2-consistent land conversion on temperature and ozone profiles and associated public health impacts in future decades.

   In addition, a sensitivity analysis will evaluate the relative effect of altering anthropogenic precursor emissions inventories upon ozone-related health impacts; altering the total population exposed to heat and/or ozone; applying different concentration-response coefficients from the epidemiological literature for temperature and ozone; changing the health-relevant threshold values; and changing the geographic focus of the risk assessment in order to look at different eastern US cities within the modeling domain, as well as zooming in on differences between individual neighborhoods within the New York Metropolitan Region. When results of a fuller set of model simulations are completed, the relative effects of these changes upon the regional health impacts projections can be more fully evaluated.

This type of integrated regional assessment allows us to project climate and thus climate-related health impacts at temporal and spatial scales relevant to local/regional planning. The risk assessment model framework for evaluating climate-related health impacts gives a high degree of transparency and adaptability to the integrated model system, which can be exported to other locales.

CONCLUSIONS

The NYCHP has developed an integrated modeling framework for assessing potential public health impacts of climate and land use changes at the regional scale. The results for temperature and ozone presented here illustrate the kinds of research and policy questions that can be explored using this framework.

An important contribution of the NYCHP to the field of integrated assessment is the application of fully down-scaled climate and air quality model outputs to health impacts projections. The resultant projections of climate change-related health impacts can be expressed at a regional- or local-scale that should prove far more useful for adaptive planning than previous projections of continental-scale climate changes. This is one of the first projects to generate county-specific projections by linking the GISS-GCM outputs to boundary condition inputs for the MM5 RCM, and subsequently linking the MM5 to CMAQ boundary conditions for ozone modeling, and estimation of public health impacts. As applied here, both the GISS-GCM and the MM5 RCM, along with the CMAQ ozone modeling, provide the advantages of:

1. daily simulations of environmental conditions, which allowed application of daily exposure-risk coefficients from the epidemiological literature on mortality and morbidity; and
2. consistent sets of emissions/growth assumptions from the IPCC SRES scenario families A2 and B2, which were used across all components of the linked impacts and whose unified assumptions gave consistency to results.

Future work will integrate land use into the fully linked modeling framework. The integrated modeling system developed here can provide a useful tool for policy makers, health officials and infrastructure planners for projecting potential impacts of climate and land use change at a level of geographic detail that is relevant to the problems they face.

NOTES

1. These projected temperatures are for the full range of 35 SRES scenarios, based on several climate models (IPCC 2001).
2. This work was supported by the US Environmental Protection Agency's National Center for Environmental Research (NCER) STAR Program, under grant R-82873301. Although the research described has been funded in part by the US Environmental Protection Agency, it has not been subjected to the Agency's required peer and policy review and therefore does not necessarily reflect the views of the Agency and no official endorsement should be inferred. Additional support was provided by grant ES09089 from the National Institute of Environmental Health Sciences (Center for Environmental Health in Northern Manhattan).

REFERENCES

Anthes, R.A. and T.T. Warner (1978), 'Development of hydrodynamic models suitable for air pollution and other mesometeorological studies', *Monthly Weather Review*, **106**: 1045–78.

Betts, A.K. (1986), 'A new convective adjustment scheme. Part I: Observational and theoretical basis', *Quarterly Journal of the Royal Meteorological Society*, **112**: 677–92.

Braga A., A. Zanobetti and J. Schwartz (2002), 'The effects of weather on respiratory and cardiovascular deaths in 12 US cities', *Environmental Health Perspectives*, **110**: 859–63.

Byun, D.W. and J.K.S. Ching (eds) (1999), *Science Algorithms of the EPA Models-3 Community Multiscale Air Quality Model (CMAQ) Modeling System* (EPA/600/R-99/030), Washington, DC: US Environmental Protection Agency, Office of Research and Development.

Carolina Environmental Programs (2003), *Sparse Matrix Operator Kernel Emission (SMOKE) Modeling System*, University of Carolina, Carolina Environmental Programs, Research Triangle Park, NC.

Clarke, K.C., S. Hoppen and L. Gaydos (1997), 'A self-modifying cellular automaton model of historical urbanization in the San Francisco Bay Area', *Environmental and Planning B*, **24**: 247–61.

Connecticut Department of Public Health (CDPH) (1998), http://www.state.ct.us/dph.

Curriero, F.C., K.S. Heiner, J.M. Samet, S. L. Zeger, L. Strug, and J.A. Patz (2002), 'Temperature and mortality in 11 cities of the eastern United States', *American Journal of Epidemiology*, **155**: 80–87.

Curriero, F.C., J.M. Samet and S.L. Zeger (2003), Re: 'On the use of Generalized additive models in time-series studies of air pollution and

health' and 'Temperature and mortality in 11 cities of the eastern United
States' (Letter to the Editor), *American Journal of Epidemiology*, **158**(1):
93–94.

Gaffin, S.R. (1998), 'World population projections for greenhouse gas
emissions scenarios', *Mitigation and Adaptation Strategies for Global
Change*, **3**: 133–70.

Gaffin, S.R. (2002), *Population projections 2000–2100 in 5-year increments*,
prepared at the request of the IPCC (personal communication);
downscaled A2 pop estimates for US by Wolfgang Lutz, IIASA, 2002, at
the request of the IPCC.

Gaza, R.S. (1998), 'Mesoscale meteorology and high ozone in the northeast
United States', *Journal of Applied Meteorology*, **37**: 961–67.

Geron, C.D., A.B. Guenther and T.E. Pierce (1994), 'An improved model for
estimating emissions of volatile organic compounds from forests in the
eastern United States', *Journal of Geophysical Research,* **99**(12): 773–91.

Gold D.R. (1992), 'Indoor air pollution', *Clinics in Chest Medicine*, **13**: 215–
29.

Hamburg, M. (1998), 'Emerging and resurgent pathogens in New York City',
Journal of Urban Health: Bulletin of the New York Academy of Medicine,
75(3): 471–79.

Hogrefe, C., J. Biswas, B. Lynn, K. Civerolo, J.-Y. Ku, J. Rosenthal, C.
Rosenzweig, R. Goldberg and P.L. Kinney (2004a), 'Simulating regional-
scale ozone climatology over the Eastern United States: Model evaluation
results', *Atmospheric Environment*, **38**(17): 2627–38.

Hogrefe, C., B. Lynn, K. Civerolo, J.-Y. Ku, J. Rosenthal, C. Rosenzweig, R.
Goldberg, S. Gaffin, K. Knowlton and P.L. Kinney (2004b), 'Simulating
changes in regional air pollution due to changes in global and regional
climate and emissions', *Journal of Geophysical Research*, **109** 2627–38.

Hong, S.-Y. and H.-L. Pan (1996), 'Nonlocal boundary layer vertical
diffusion in a medium-range forecast model', *Monthly Weather Review*,
124(2): 322–39.

Intergovernmental Panel on Climate Change (IPCC) (2001), *Working Group I
Third Assessment Report*, Cambridge, UK: Cambridge University Press.

Kalkstein, L.S. and J.S. Greene (1997), 'An evaluation of climate/mortality
relationships in large US cities and the possible impacts of a climate
change', *Environmental Health Perspectives*, **105**: 84–93.

Katsouyanni, K., A. Pantazopoulou and G. Touloumi (1993), 'Evidence for
interaction between air pollution and high temperature in the causation of
excess mortality', *Archives of Environmental Health*, **48**: 235–42.

Kinney, P.L. (1999), 'The pulmonary effects of outdoor ozone and particle air
pollution', *Seminars in Respiratory and Critical Care Medicine*, **20**: 601–
7.

Kinney, P.L., M. Aggarwal, M.E. Northridge, N.A.H. Janssen and P. Shepard (2000), 'Sidewalk exposures to $PM_{2.5}$ and diesel exhaust particles in Harlem', *Environmental Health Perspectives*, **108**: 213–18.

Klinenberg, E. (2002), *Heatwave: Social Autopsy of Disaster in Chicago*, Chicago, IL: University of Chicago Press.

Knowlton, K., C. Hogrefe, J.E. Rosenthal, B. Lynn, C. Rosenzweig, S. Gaffin, R. Goldberg, R. Civerolo, J-Y. Ku and P.L. Kinney (2004), 'Climate-related changes in ozone mortality over the next 50 years in the New York City metropolitan region', *Environmental Health Perspectives*, **112**(15): 1557–63.

Liu, J., G.C. Daily, P.R. Ehrlich and G.W. Luck (2003), 'Effect of household dynamics on resource consumption and biodiversity', *Nature*, **421**: 530–33.

Lynn, B.H., L. Druyan, C. Hogrefe, J. Dudhia, C. Rosenzweig, R. Goldberg, D. Rind, R. Healy, J. Rosenthal and P. Kinney (2004), 'Sensitivity of present and future surface temperatures to precipitation characteristics', *Climate Research*, **28**:53–65.

Mannino D.M., D.M. Homa, C.A. Pertowski, A. Ashizawa, L.L. Nixon, C.A. Johnson, L. Ball, E. Jack and D.S. Kang (1998), *Surveillance for Asthma United States, 1960–1995*, CDC MMWR, **47**(SS-1): 1–28.

Mlawer, E.J., S.J. Taubman, P.D. Brown, M.J. Iacono and S.A.Clough (1997), 'Radiative transfer for inhomogeneous atmosphere: RRTM, a validated correlated-k model for the longwave, *Journal of Geophysical Research*, **102**: 16 663–82.

Nakićenović, N., J. Alcamo, G. Davis, B. de Vries, J. Fenhann, S. Gaffin, K. Gregory, A. Grübler, T.Y. Jung, T. Kram, E. Lebre La Rovere, L. Michaelis, S. Mori, T. Morita, W. Pepper, H. Pitcher, L. Price, K. Riahi, A. Roehrl, H-H. Rogner, A. Sankovski, M. Schlesinger, P. Shukla, S. Smith, R. Swart, S. van Rooijen, N. Victor and Z. Dadi (2000), *IPCC Special Report on Emissions Scenarios (SRES)*, *Working Group III, Intergovernmental Panel on Climate Change*, Cambridge: Cambridge University Press, http://www.grida.no/climate/ipcc/emission.

National Academy of Sciences (NAS) (2000), *Clearing the Air: Asthma and the Indoor Environment*, Washington: National Academy Press.

National Oceanic and Atmospheric Administration (NOAA) (2003), *Natural Hazard Statistics: Weather Fatalities*, National Weather Service, Office of Climate, Water and Weather Service, last updated 13 November, 2003, http://www.nws.noaa.gov/om/hazstats.shtml.

New York City Department of Health and Mental Hygeine (NYC DOH) (2003), 'NYC vital signs: Asthma can be controlled', **2**(4), April.

New York City Department of Health and Mental Hygeine (NYC DOH) (1998), 'Summary of reportable diseases and conditions', C. Lowe (ed.), *City Health Information* **17**(1): 1–28.

Northridge, M.E., J. Yankura, P.L. Kinney, R.M. Santella, P. Shepard, Y. Riojas, M. Aggarwal and P. Strickland (1999), 'Diesel exhaust exposure among adolescents in Harlem: A community-driven study', *American Journal of Public Health*, **89**(7): 998–1002.

Regional Planning Association (RPA) (1996), *A Region at Risk:A Summary of the Third Regional Plan for the New York-New Jersey-Connecticut Metropolitan Area*, New York City: RPA, http://www.rpa.org.

Russell, G.L, J.R. Miller and D. Rind (2003), 'A coupled atmosphere-ocean model for transient climate change studies', *Atmosphere-Ocean*, **33**(4): 683–730.

Schwartz, J. and D. Dockery (1992), 'Increased mortality in Philadelphia associated with daily air pollution concentrations', *American Review of Respiratory Diseases*, **142**: 600–604.

Seaman, N.L. and S.A. Michelson (2000), 'Mesoscale structure of a high-ozone episode during the 1995 NARSTO-Northeast study', *Journal of Applied Meteorology*, **39**: 384–98.

Sillman, S. and P.J. Samson (1995), 'Impact of temperature on oxidant photochemistry in urban, polluted, rural and remote environments', *Journal of Geophysical Research*, **100**(14): 175–88.

Thurston, G.D. and K. Ito (1999), 'Epidemiological studies of ozone exposure effects', in J. Ayers, R.L. Maynard and R. Richards (eds), *Air Pollution and Health*, San Diego, CA: Academic Press, pp. 485–510.

Thurston, G.D. and K. Ito (2001), 'Epidemiological studies of acute ozone exposures and mortality', *Journal of Exposure Analysis and Environmental Epidemiology.*, **11**: 286–94.

US Centers for Disease Control and Prevention (US CDC) (2004), CDC Wonder, Dept of Health and Human Services, Epidemiology Program Office, Division of Public Health Surveillance and Informatics, Accessed 19 January, 2004, http://wonder.cdc.gov/mortSQL.html.

US Environmental Protection Agency (US EPA) (1994), *User's Guide to Mobile5 (Mobile source emission factor model)* (EPA/AA/TEB/94/01), Ann Arbor, MI.

US Environmental Protection Agency (US EPA) (2003a), *National Air Quality and Emissions Trends Report, 2003 Special Studies Edition* (EPA 454/R-03-005), Research Triangle Park, NC, September 2003.

US Environmental Protection Agency (US EPA) (2003b), *Latest Findings on National Air Quality: 2002 Status and Trends* (EPA 454/K-03-001), Research Triangle Park, NC, August.

US Environmental Protection Agency (US EPA) (2003c), *Environmental Protection Agency, Clearinghouse for Inventories and Emissions Factors, National Emissions Inventory*, Accessed March 2003, http://www.epa.gov/ttn/chief/trends/index.html.

US Environmental Protection Agency (US EPA) (2004), *US EPA Green Book: Nonattainment Areas for Criteria Pollutants*, last updated 20 January, 2004, http://www.epa.gov/oar/oaqps/greenbk.

Williams, E.J., A. Guenther and F.C. Fehsenfeld (1992), 'An inventory of nitric oxide emissions from soils in the United States', *Journal of Geophysical Research*, **97**(D7): 7511–19.

7 Climate's Long-term Impacts on Urban Infrastructures and Services: The Case of Metro Boston

P. Kirshen, M. Ruth and W. Anderson

INTRODUCTION

Infrastructure provides human, environmental and economic services and directly contributes to the quality of life. Services typically include flood control, water supply, drainage, wastewater management, solid and hazardous waste management, energy, transportation, constructed facilities for residential, commercial, and industrial activities, communication and recreation.

The American Society of Civil Engineers (ASCE 1998) estimates that in 1992 the total value of US infrastructure investments was approximately 40 percent of GDP and in 1982 was as high as 55 percent of GDP. However, as significant as these values are, they barely begin to reflect the true value of infrastructure. Without infrastructure the US economy could not function. Many human and environmental systems would collapse. Quality of life would be poor.

Since most human and economic activities in the US are and will be associated with areas of high population densities, metropolitan infrastructure systems and the services they provide are particularly important in order to achieve and maintain a high quality of life. Yet even though urban infrastructure systems are important and are designed according to socioeconomic and environmental conditions that are very sensitive to climate, there have been no major integrated assessments of the impacts of climate change on metropolitan infrastructure systems and services in the US.

The Climate's Long-term Impacts on Metro Boston (CLIMB) project conducted from 1999 to 2004 explores potential changes in infrastructure systems and services in Metro Boston in response to changes in climate, socioeconomic and technological developments. Potential changes play themselves out across space and time, owing to differences in infrastructure densities and use, differences in environmental conditions, and the long-term nature of climate change as well as the long-lived nature of the various infrastructure systems. To capture spatial variations in climate change impacts on Metro Boston, seven sub-regions or zones are distinguished (Figure 7.1) such that:

1. coastal regions are treated separately from regions inland;
2. areas north of the city of Boston, which have different coastal properties and socioeconomic features, are delineated from southern parts of the Metropolitan Area Planning Council (MAPC) region;
3. highly urbanized areas are dealt with separately from suburbs; and
4. rapidly growing suburbs are distinguished from already highly developed and densely populated ones.

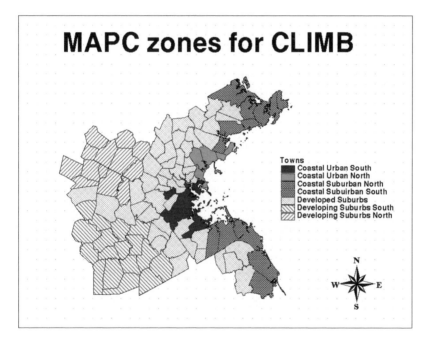

Figure 7.1   Massachusetts and Metropolitan Boston with seven zones defined for CLIMB

The following sections describe the project structure and approach of CLIMB, basic scenario assumptions and analyses of specific infrastructure systems and services, including water supply, water quality, flood control, transportation, energy and public health. The chapter closes with an integrated analysis and a summary and conclusions.

PROJECT STRUCTURE AND APPROACH

The complexity that arises from climate–infrastructure–economy–society interactions in space and time can easily overwhelm the ability of decision makers to systematically explore the consequences of alternative actions. CLIMB provides tools for scenario analysis and decision support to assess potential impacts of climate change on individual infrastructure systems and services (water supply, water quality, flood control, surface transportation, energy, public health), as well as their integrated impacts. Each of the infrastructure systems are described by a set of state variables, pressures that are exerted on the status of the system, impacts that those pressures have on system performance, and potential response strategies to affect system performance as laid out in Figure 7.2. A set of indicators are developed to assess the ability of each individual infrastructure system to provide services over time, and the impacts that climate change and other drivers have on individual infrastructure systems and their collective performance.

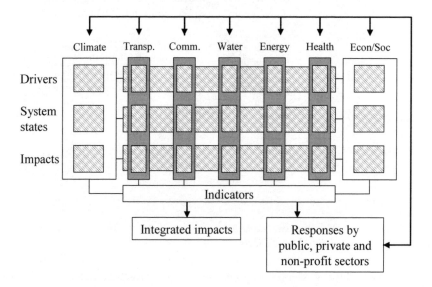

Figure 7.2　　Interrelationships among system components

The CLIMB project consists of infrastructure-specific and cross-cutting analyses that are carried out in close collaboration with stakeholders in the region. CLIMB has made stakeholders an integral part of the project for the following reasons:

There is an inherent, democratic value in including in a research project those segments of society who are ultimately affected by the decisions that are based, at least in part, on that research. The responsibility of researchers to include stakeholder involvement is the greater the more a project attempts to address issues that are of direct relevance to the lives of stakeholders, and the more the project relies on funds made available by the tax payers.

Stakeholders can possess valuable knowledge that may be difficult to access by researchers. Stakeholder involvement can thus not only broaden the information base on which science operates, but can provide a powerful means for 'ground-truthing' of data, models and model scenarios.

Different institutions typically have taken it upon themselves to be advocates for individual segments of society, economy or environment. By including in science and decision making stakeholders from a range of social, cultural and economic institutions, alliances can be forged to help disseminate research results, and help the broader public make the connection between a project and people's own personal and professional lives. Leveraging the interest of institutions in disseminating select pieces of information is particularly valuable in projects that tackle highly complex human–environment systems in interaction. For example, the topic of climate change can become less abstract and more tangible to individuals if implications for people's economic welfare or personal health can be identified, and if decisions of individual firms and consumers can be influenced.

SCENARIOS

The impacts of the present climate and different changing climates are simulated annually for the period 2000 to 2100 for infrastructure systems and services under three different, internally consistent socioeconomic and technological scenarios. Since climate conditions vary from year to year even without climate change, different sequences of yearly climate conditions are simulated for each separate climate change scenario. If sufficient sequences are explored, then the aggregate output of all these simulations is the expected impact under a particular climate change scenario given a socioeconomic and technology scenario.

Socioeconomic and Technological Scenarios

Socioeconomic scenarios are created from population, household and economic forecasts from regional and national sources (MAPC 2000; NPA Data Services Inc. 1999a,b; 2001a,b) using the US Bureau of Labor Statistics and the US Census Bureau historical data as benchmarks to adjust projection series. These forecasts are available at the community level until 2025 and are used to down-scale the corresponding socioeconomic variables available from the New England Climate Assessment, which contains scenarios until 2050. After 2050, we assume no further changes. The demographic variables used in the CLIMB project are resident population and employment in the retail, basic, and industrial economic sectors. Technology scenarios assume diffusion of various advanced engineering practices in the region. The technology scenarios are thus closely related to behavioral and policy variables (Table 7.1), creating the challenge of making all exogenous drivers consistent with each other across space and time.

As the various exogenous drivers influence the internal dynamics of infrastructure systems – for example, their reliability in providing services to the region – different investment and policy responses may occur (Table 7.1). The Ride It Out scenario in essence assumes that no adaptation to climate change occurs and that damages and benefits continue to occur with no attempts by society to minimize damages or maximize benefits. In contrast, the Green scenario assumes conscious, sustainable responses to observed trends, as well as pro-active or anticipatory implementation of policies and technologies in efforts to counteract, and prepare for, adverse climate impacts. Some of the practices might be put in place before impacts are felt (for example, moving occupants out of flood plains), after impacts occur, or at the end of lifecycles of infrastructure systems. The Build Your Way Out scenario assumes that replacement of failed systems is undertaken and susceptible systems are protected by structural measures. For example, in the case of sea level rise, the Ride it Out scenario would be characterized by a lack of attempts to protect property and just rebuild after damages. In contrast, Build Your Way Out adaptation responses may include a set of engineering measures that help to protect against it (for example holdback the sea with structures such as seawalls, barriers or pumps; or foster beach nourishment).

Regional Climate Scenarios

For each year of the historic climate records at Boston's Logan Airport (the longest, most consistent set in Metro Boston), we have determined the values of the climate parameters that impact each system except for water supply. Because the streamflow for the regional water supply system that serves much of Metro Boston comes from regions outside of Metro Boston, we have used meteorological data from those regions for water supply. Data for the two water supply sources are from the United Kingdom LINK dataset (Viner 2000). The climate dataset was organized with software developed for the CLIMB project by Horwitz (2004).

Table 7.1   *Three CLIMB model scenarios*

Policy	Demographic	Economic	Technology
	Ride It Out		
Present trends in region continue. There are no adaptation actions.	Same as current MAPC scenario of continued sprawl, low population growth rate, major growth at fringes and outside of region.	Same employment by sector as current MAPC and New England Regional Assessment scenarios.	Low rate of penetration of green and innovative technology by sector.
	Green		
Restrictions on construction locations. Stronger building codes. Natural hazard zoning. Emphasis on more centralized development.	Same population growth as Ride It Out	Same as Ride It Out	Higher rate of green technology penetration than Ride It Out.
	Build Your Way Out		
Same as Ride It Out but replace and protect systems as they fail.	Same as Ride It Out	Same as Ride It Out	Same as Ride It Out

Building upon this data set, we modeled responses of infrastructure systems to the climate for every year over the period 2000 to 2100. To develop many possible representative time-series of the present climate and to extend our climate record from the present length of approximately 50 years to 100 years, we used moving block bootstrapping (Vogel and Shallcross 1996). This is a non-parametric statistical method that maintains probability relationships of time-series values both within years and over years and consists of sampling with replacement from the existing time-series of annual climate events until the desired number of time-series of desired length is obtained. To model time-series of climate change scenarios, trends of climate changes were applied to the set of time-series representing the present climate.

For example, if sea level rose by 1 percent per year, each year the sea level of each time-series representing the present climate was increased by 1 percent. This set of changed climate time-series can then be used to explore the impacts of changing climate in Metro Boston using the Monte Carlo simulation.

In most analyses, we used two Generalized Circulation Model (GCM) climate scenarios and various sensitivity analyses. The GCM scenarios were the same Canadian Climate Center (CCC) and Hadley Center scenarios that were used in New England during the recent US national assessment of climate change (New England Regional Assessment Group 2001). The atmosphere–ocean general circulation models used for the scenarios are the Canadian CGCM1 and the Hadley HadCM2. The greenhouse gas emission scenario assumed a 1 percent annual increase in equivalent CO_2 and included the direct effects of sulfate aerosols in the atmosphere (IS92a scenario). Scenario data were obtained for the inland grid cell closest to our study area for 2030 and 2100 climate scenarios. A summary of scenario impacts in our study area is in Table 7.2. One scenario is humid and warm, the other more humid and less warm.

Table 7.2   Summary 2100 annual climate change scenarios for Metro Boston

GCM	Temperature increase (°C)	Precipitation increase (%)
CGCM1	4.8	5.9
HadCM2	2.95	23.0

WATER SUPPLY AND DEMAND

Water Supply

Metro Boston is served by three types of water supply systems – systems that are totally self supplied by local ground and surface water systems; systems supplied by a large regional reservoir-based water supply system located outside and west of Metro Boston operated by the Massachusetts Water Resources Authority (MWRA); and systems partially supplied by both local sources and the MWRA. The MWRA supplies most of the water to the region, drawing on surface water from the central portion of the state and delivering it to the region as well as some other users outside the region. The CLIMB zones were divided into sub-areas corresponding to the three types of systems within each zone.

The water supply model determines the impacts of climate and other changes upon water supply in each sub-area by initially calculating annual demand for water by each of four sectors (domestic, retail, service and industry) to determine the total demand. Water supply availability for that year from each sub-area and from MWRA reservoirs are determined based upon water supply infrastructure and policy constraints, past reservoir and groundwater storage, and that year's climatic influences upon streamflows and groundwater recharge. The available sub-area supply after supplying low flow requirements is then compared to the sub-area demand. If there is a shortage in a partially supplied system, the present policy is to attempt to meet the shortage from the regional system. If there are supply shortages from the regional system, each supplied system receives the same percentage of reduced demand. Local systems presently must respond without assistance to local shortages.

Water Demand

Water demand for domestic and each employment sector were calculated as the product of the unit demand and total number of units given by demographic and employment projections. Due to a lack of adequate information, unit demand quantities were assumed to remain constant at their 2000 values: 0.3440 cubic meters/person/day (cmpd) (90 gallons/day/person) for domestic demand; 0.0568 cmpd (15 gallons/day/employee) for basic employment; 0.3503 cmpd (92.6 gallons/day/employee) for retail employment; and 0.0855 cmpd (22.5 gallons/day/employee) for service employment.

An annual requirement for the quantity of water that must remain in each locally and partially supplied sub-area after water consumption was set as the estimated water in each sub-area that would remain in the sub-area after Year 2000 consumptive losses and outflows of wastewater in sewers under low flow water availability conditions that occur with a probability of 10 percent, or the 10-year low flow. During the modeling, each year the model determined the quantity of water that would remain in a sub-area given consumption and sewer outflows and then adjusted the total demand if necessary to maintain the target amount of water to remain in each sub-area.

Water withdrawals by self- and partially-supplied systems were also limited by the Massachusetts Water Management Act (WMA) limits on each municipality. Values were set at the maximum amounts presently permitted over the next 15 years.

Hydrologic Modeling

The 'abcd' water balance model (Thomas 1981; Fernandez et al. 2000) was used to estimate the annual surface and groundwater available in each self-

and partially-supplied sub-area and the MWRA system for each year of the simulation period. It is a non-linear watershed model that accepts precipitation and potential evapotranspiration (PET) as input and produces streamflow as output; internally it represents soil moisture storage, groundwater storage, direct runoff and actual evapotranspiration. The model was calibrated for two relatively non-anthropogenically altered streamflow time-series outside of Metro Boston and each sub-area was then assigned to one of the representative time-series, referred to as North and South. The area of each sub-area was then multiplied by the appropriate annual unit runoff to determine the total water availability for any year. Estimating streamflows entering the MWRA reservoir system required modeling the Ware and Connecticut Rivers.

Monthly time-series of precipitation, vapor pressure, cloud cover and mean temperature were obtained from the LINK database (Viner 2000) for two grid centers 71.25W, 42.75N and 72.25W, 42.75N. Monthly time-series of potential evapotranspiration were estimated using Priestley–Taylor method for the period 1901–1995 (Shuttleworth 1993). The monthly time-series of potential evapotranspiration and precipitation at each grid center were aggregated to produce annual time-series of precipitation and potential evapotranspiration.

The streamflow data available for the North and South zones and the Ware River were calibrated with the annual time-series of precipitation and potential evapotranspiration of the grid center at 71.25W, 42.75N, whereas the calibration for the Connecticut River was carried out with the climate data available at the grid center 72.25W, 42.75N. All the calibrations were carried out in an EXCEL worksheet by reducing the sum of squares of errors and by using the genetic algorithm based non-linear optimization. Table 7.3 gives the summary statistics of the calibrated flows and the observed flows. The calibrations are strong and show similar elasticities of streamflow to precipitation as derived by Sankarasubramanian and Vogel (2001) for the region.

Table 7.3   Summary statistics of the observed and calibrated flows and the model parameters

Zone	Observed mean (mm/yr)	Calibrated mean (mm/yr)	Observed std. dev. (mm/yr)	Calibrated std. dev. (mm/yr)	Coefficient of determination
Ware	600.88	601.57	158.07	139.51	0.747
Conn.	611.67	611.33	122.24	109.77	0.694
North	556.74	557.91	153.77	139.28	0.740
South	669.59	671.00	187.13	160.14	0.789

The safe yield of the MWRA system was simulated for the period 1930–79 (Kirshen and Fennessey 1995). It was found that the safe yield is approximately 400.62 million cubic meters (MCM), or 290 million gallons per day (MGD), as failures started to occur at 404.79 MCM (or 292 MGD). This agrees well with the MWRA safe yield estimate of 414.46 MCM (or 300 MGD).

CLIMATE CHANGE SCENARIOS

Using data for potential climate change scenarios from the Canadian Climate Center and the Hadley Center described above, changes in annual PET were calculated. In addition, it was assumed that the albedo changed gradually from 0.15 in 2000 to 0.12 in 2100 to account for more urbanization and less snow. There were no adjustments made in changes in PET due to an enriched CO_2 atmosphere.

In order to model possible increases in the variability of annual water availability in the future due to climate change, the precipitation in each year for each of the ten bootstrapped time-series was represented as the sum of the mean precipitation over the period and a fraction of the standard deviation over the period. The standard deviation was then constantly increased each year from 2000 so that by 2100 it was 30 percent more than the deviation of the original time-series and a new precipitation time-series determined as the sum of the mean and an increased fraction of the original standard deviation. This procedure maintained the annual mean but modeled a flow of increased annual variability. The mean annual precipitation time-series used in the model were then the ten bootstrapped time-series with the standard deviation increased constantly. The mean annual streamflow changes for the years 2030 and 2100 are in Table 7.4.

Table 7.4   Climate and streamflow scenarios 2030 and 2100

	Canadian Climate Center	Hadley Center
Precip. 2030/Precip. 2000	1.002	1.102
PET 2030/Precip. 2000	1.085	1.066
Runoff 2030/2000	0.953	1.126
Precip. 2100/Precip. 2000	1.059	1.230
PET 2100/PET 2000	1.145	1.078
Runoff 2100/2000	1.010	1.333

Adaptation Scenarios

The same types of adaptation scenarios as described above were used to define a Base Case (present climate and growing demands due to demographic changes), a Ride It Out scenario (changed climate, same demands as Base Case, and increased low flow requirement after 2020; 20 percent more environmental flows in local and regional systems than now after allowing for consumptive losses and return flows to sewer systems as assumed policy response to attempt to mitigate some of the negative environmental consequences of climate change upon water quality), a Green scenario (same as Ride It Out except decrease demands to 65 percent of present demands after 2025 to attempt to mitigate impacts by demand management), and a Build Your Way Out scenario (same as Ride It Out except attempt to supply all local self-supply deficits from MWRA regional system).

Analyses were only carried out for the changes in streamflow under the Canadian climate change scenario because the large increases in water availability under the Hadley scenario appear to be sufficient to meet any demand increases in the future. Thus the Canadian climate change scenario represents a possible lower bound on negative impacts climate.

As can be seen from Tables 7.5 and 7.6, average reliabilities decrease over time. During the period 2000–2050 the communities relying totally or partially upon the MWRA system in all scenarios are approximately 100 percent reliable. Under the Base Case, Ride It Out, and Green scenarios, local reliabilities are less and when a shortage occurs, it is at least 75 percent of the demand. Under Build Your Way Out, local systems are approximately 100 percent reliable. As can be seen in Table 7.6 during the period 2000–2100 in the Base Case, Ride It Out and Green scenarios, the communities totally or partially supplied by the MWRA are 100 percent reliable. It is not shown but the local reliabilities over the period 2050–2100 are lower than during the period 2000–2050; approximately 80 percent. Under Build Your Way Out, the reliability of the MWRA system decreases to a still probably acceptable level of 0.97 and the average reliability of the local systems improve to that value. The average shortage is small.

If a reliability of 0.90 can be considered marginally acceptable for a water supply system (the desired reliability is actually closer to 1.00) because a manager can respond to shortages by ordering significant conservation, the results show that there are significant decreases in the reliabilities of locally supplied systems after 2050 in the Ride It Out and Green scenarios when the reliabilities decrease to 0.80. The shortages are less under the Green scenario. Under the adaptation actions considered in the report, only by the local systems using the regional MWRA system to supplement their supplies is it possible to maintain acceptable local water supplies under climate and demographic changes. Even with the higher demands on it under Build Your Way Out, the reliability of the regional MWRA systems remains manageable in the future under climate and demographic changes.

Table 7.5   2000–2050 average reliabilities and deficits

MWRA reliability	Average deficit as ratio of average annual demand	Average regional shortage (MCM)	Local supply average reliability	Average deficit as ratio of average demand
		Baseline		
1.00	0.00	0.00	0.93	0.75
		Ride It Out		
1.00	0.00	0.00	0.86	0.84
		Green		
1.00	0.00	0.00	0.87	0.79
		Build Your Way Out		
~1.00	0.05	30.75	~1.00	0.05

Table 7.6   2000–2100 average reliabilities and deficits

MWRA reliability	Average deficit as ratio of average demand	Average regional shortage (MCM)	Local supply average reliability	Average deficit as ratio of average demand
		Baseline		
1.00	0.00	0.00	0.93	0.78
		Ride It Out		
~1.00	0.05	20.83	0.82	0.86
		Green		
1.00	0.00	0.00	0.83	0.84
		Build Your Way Out		
0.97	0.33	197.26	0.97	0.33

Presently, the MWRA is not obligated to serve locally supplied systems in event of temporary or permanent shortages. Therefore, local systems could consider anticipating climate and demographic changes by managing demand to minimize shortages, increasing in-stream flows, increasing system storage capacity though reservoirs or aquifer use, or considering using such water supply sources as reclaimed wastewater and desalination.

WATER QUALITY

Introduction

Dissolved oxygen (DO) is an important indicator of water quality – essentially all animals in rivers require oxygen (Hemond and Fechner 1994). Insufficient DO in rivers and streams is caused by the consumption of oxygen during decomposition of organic matter and respiration by plants and animals exceeding the rate at which oxygen enters the water body. Oxygen enters by its transfer from the atmosphere (the process of reaeration) and by plant photosynthesis (Chapra 1997). Sources of organic matter in streams include sewage from wastewater treatment plants, dead plants, non-point source (NPS) runoff and bottom sediments. Besides direct inputs of organic matter, DO shortages are also exacerbated by large inputs of nutrients from many of the sources above that cause excessive plant growth or eutrophication. Low flows can result in decreased DO because of decreased dilution of pollution and, in some cases, reduced reaeration. High temperatures decrease the maximum possible DO in rivers (the saturation amount) and hasten decomposition and plant growth. On a daily basis, minimum DO usually occurs in the early morning because plants and animals have been respiring during the night without photosynthesis.

DO levels below 5 mg/L may impair the functioning and survival of aquatic communities, and levels below 2 mg/L may lead to the death of some fish species. Low DO can also lead to the release of toxic metals and phosphorus from bottom sediments (Chapra 1997).

Population growth typically results in an increased amount of organic matter entering rivers because of more sewage and more NPS loads. Larger populations may cause lower streamflows because of more water consumption. NPS loads may also increase under climate change because annual precipitation and the frequency of extremely high precipitation events may increase in some regions of the US (Meehl et al. 2000).

As an example of possible integrated long-term climate and socioeconomic impacts on water quality, we examine the integrated impacts in a watershed in Metropolitan Boston, choosing the Assabet River as a case study. The river flows in a northeasterly direction through several highly populated municipalities in Metropolitan Boston before merging with the Sudbury River to form the Concord River. It drains a 458-square kilometer watershed, with a 2000 population of 177 000 and an average annual rainfall of approximately 100 centimeters. Predominant land uses in the watershed are 35 percent urban, 6 percent agricultural and 50 percent forested (MASSGIS 2000). The river drops 52 meters in elevation along its length, with the steepest drops coming just after each of the five major impoundment dams. The river is swampy, slow-moving and relatively narrow and shallow, typically 10 to 20 meters wide and 0.5 to 1.2 meters deep except behind the dams (ENSR International 2000). In summer it has high water temperatures and is exposed to long hours of daylight. Four major municipal wastewater treatment plants (WWTP's), located in Westborough, Marlborough, Hudson, and Maynard, discharge to the river. Low DO and eutrophication due to point and non-point sources are presently identified as the major water quality problems of the basin. Based upon field surveys, ENSR International (2001) reports that the eutrophication results from excess nutrients entering the river primarily from the wastewater treatment plants. Non-point and sediment sources are minor annual and summertime contributors to the nutrient loads. The major causes of low DO are plant respiration and sediment oxygen demand (SOD). The organic loads from the WWTPs are not major causes of the low DO. Management of DO in the northeast has historically been done under late summer flow conditions when flows are lowest and warmest. The design flow is the 7 day average of low flows that has a recurrence interval of 10 years ($_7Q_{10}$).

Table 7.7 shows estimated natural and anthropogenic components of the flow measured or estimated at three stations in the river during $_7Q_{10}$ conditions. Stations 1 and 2 are in the upper portions of the basin and Station 3 is further downstream. The flow distribution is based upon an estimate of natural $_7Q_{10}$ flows at the stations using the USGS web-based program Streamstats (USGS 2001) and measured flows at the stations on 21 August 1999, when conditions approximating natural $_7Q_{10}$ conditions occurred (US EPA 2000). Anthropogenic flows increase upstream of Stations 1 and between Stations 1 and 2, but decrease between Stations 2 and 3. Upstream of Station 2 there are large wastewater discharges, which in some cases are from water supply imported from outside the basin. Downstream of Station 2, the local withdrawals and consumption exceed discharges from wastewater treatment plants and local septic systems.

Table 7.7   *Measured and estimated river discharges in August 1999*

Flow (m^3/s)	Station 1	Station 2	Station 3
Measured Total	0.147	0.423	0.329
Natural Component	0.022	0.086	0.216
Anthropogenic Component	0.125	0.337	0.113

Water Quality Model

The model of the DO of the Assabet river system simulates steady-state conditions for loads, flows and other key variables during $_7Q_{10}$ conditions. The diurnal impacts of solar radiation on plant respiration, photosynthesis and water temperature, however, were modeled explicitly. The formulation is similar to QUAL2E (described in Chapra 1997).

The river was divided into 20 reaches or control volumes with entering flows, loads and water withdrawal distributed among the reached. At the end of each river reach, the concentrations of biochemical oxygen demand (BOD), nutrients and dissolved oxygen were determined by dynamic mass balance equations. The average depth and velocity of each reach were calculated using stage-discharge relationships for each reach (Wilson 2002). To capture the impact of organic matter entering in runoff during the other, non-modeled, seasons a prescribed SOD value was used in the model.

Figure 7.3 shows calibration of the model for 24 July 1999. The spatial and temporal variation of the dissolved oxygen in the river is captured reasonably well given the inherent sampling error in the data and the steady-state formulation. DO deficits are greatest just upstream of dams.

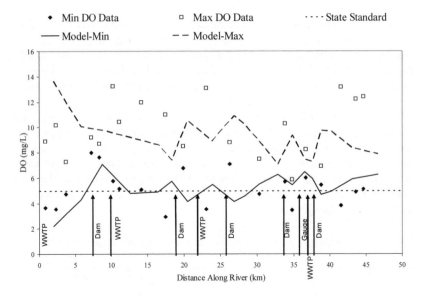

Figure 7.3   Assabet River model calibration (24 July 1999)

Climate Change and Flow Scenarios

Analyses of possible changes in the DO of the Assabet River were conducted under the climate, population, streamflow and load scenarios in Tables 7.8 and 7.9. The base case scenario is for the present point and non- point source loads, the present $_7Q_{10}$ flow and the atmospheric conditions that result in a water temperature of 26.7°C, which is the present design water temperature for DO. We found the historic combination of air temperature (30.55°C), dew point temperature (20.55°C), relative humidity (65 percent), wind speed and cloud cover values for Boston produced a water temperature of 26.7°C under $_7Q_{10}$ flow conditions. Because sensitivity analysis indicated that DO is not sensitive to wind speed and cloud cover within reasonable ranges, they were kept constant during the later analyses.

For each scenario, Table 7.9 reports the cumulative sum of the products of the DO deficit in each reach times the length of the reach (Prior Deficit, referred to as overall water quality). As can be seen, in all cases the DO deficit is larger in almost all reaches under the changed climate and population change scenarios.

The analysis also shows the sensitivity of the system to air temperature where a 2.4°C increase in air temperature results in a DO decrease of at least 0.5 mg/l along most of the river. The analysis also shows the importance of flow in the system where in many cases a change in flow compensated for an increase in loads. Comparing the Base Case and Scenario 1, a 20 percent increase in total flow in the upper reaches essentially resulted in no changes in DO in the upper reaches even though point source loads there increased 20 percent. As expected, flow decreases lowered water quality as seen in comparing the downstream impacts of the Base Case and Scenario 1. The Hadley Center scenarios resulted in increased annual non-point source loads and higher $_7Q_{10}$ flows than the Canadian Climate Center scenarios. Comparing the results of the scenarios in 2050 when the air temperatures of both are similar shows that the impacts of annual non-point source load increases are significant. The overall water quality of the Hadley Center scenario is worst even though it has higher flows; the flows river cannot compensate for higher non-point source loads (Scenarios 2 and 6). The impacts of the Canadian Climate Center and the Hadley Center scenarios for 2050 and 2100 are worse than the impacts of population growth alone in the region (Scenarios 1, 2, 3, 6 and 7). As expected, combined climate and socioeconomic changes result in the greatest impacts; DO is lower by 1 mg/l to over 2 mg/l in river reaches compared to the base case. This is because growth in this region not only increases loads but also decreases flows in downstream areas of river due to water supply increases.

Adaptation

We only examined in detail the adaptation option of decreasing the pollution loads entering the river from point and non-point sources. We assumed that the costs to adapt to the climate and population changes were the extra treatment costs to meet the water quality standard of 5.0 mg/l. Costs included control of both point and non-point source loads to the river. In our analyses

Table 7.8   *Summary of modeling scenarios*

Scenario number	Scenario description	Air temperature (°C)	Δ Mean annual precipitation (ratio)	Δ $_7Q_{10}$ and low flow tributary loads (ratio)
B	Base case	30.60	1.000	~1.00
	No climate change, no demand management, 2050 population increase	30.60	1.000	~1.00
2	2050 CC climate change, no population increase	33.00	1.018	~1.0
3	2100 CC climate change, no population increase	35.07	1.059	~1.0
4	2050 CC climate change, 2050 population increase	33.00	1.018	~1.0
5	2100 CC climate change, 2050 population increase	35.07	1.059	~1.0
6	2050 HC climate change, no population increase	33.41	1.140	1.24
7	2100 HC climate change, no population increase	33.90	1.230	1.45
8	2050 HC climate change, 2050 population increase	33.41	1.140	1.24
9	2100 HC climate change, 2050 population increase	33.90	1.230	1.45
10	2050 CC climate change, decrease in $_7Q_{10}$, 2050 population increase	33.00	1.018	0.90
11	2100 CC climate change, decrease in $_7Q_{10}$, 2100 population increase	35.07	1.059	0.80
12	2050 CC climate change, decrease in $_7Q_{10}$, 2050 population increase, demand management resulting in no change in anthropogenic flows	33.00	1.018	0.90
13	2100 CC climate change, 2100 population increase, decrease in $_7Q_{10}$, demand management resulting in no change in anthropogenic flows	35.07	1.059	0.80

Note: HC = Hadley Center Climate Scenario; CC = Canadian Climate Center Scenario

Table 7.9   *Scenario load and anthropogenic flow changes with resulting DO deficit impacts*

Scenario number	Change in mean annual flow and annual nps load and sod (ratio)	Average change in anthropogenic flows (ratio), upper/lower reaches	Change in point source load (ratio), upper/lower reaches	Prior cumulative do-deficit (mg/L-km)
B	1.00	0.00	1.00	40.4
1	1.00	1.20/0.90	1.20/1.35	52.1
2	1.00	1.00/1.00	1.00/1.00	65.1
3	1.04	1.00/1.00	1.00/1.00	76.6
4	1.00	1.20/0.90	1.20/1.35	79.1
5	1.04	1.20/0.90	1.20/1.35	92.1
6	1.24	1.00/1.00	1.00/1.00	75.8
7	1.40	1.00/1.00	1.00/1.00	81.7
8	1.24	1.20/0.90	1.20/1.35	89.3
9	1.40	1.20/0.90	1.20/1.35	94.1
10	1.00	1.20/0.90	1.20/1.35	81.9
11	1.04	1.30/1.10	1.30/1.40	98.2
12	1.00	1.00/1.00	1.20/1.35	81.7
13	1.04	1.00/1.00	1.30/1.40	100.7

with NPS management, we did not include the probable increase in low flows due to increased recharge of stormwater with NPS management practices. If these increased $_7Q_{10}$ by 10 or 20 percent, we would not expect the impacts on treatment costs to be large because natural low flows are a small proportion of the total flow in the river.

Point source load reduction
Point load reduction cost was calculated based upon the capital cost of adding an ultra-filtration unit at each of the four WWTPs with ferric chloride addition prior to filtration and operation and maintenance costs. We also added 18 percent of the construction cost to include design and construction engineering, and fiscal, legal and administrative costs (Town of Holliston 2002). Operation and maintenance were based upon electrical and chemical usage by the plant. We did not include costs of additional sludge removal from the ultra-filtration plants. Electrical usage was estimated to be $0.11 kwh/m^3 of discharge (Ionics, Inc. 2000). We used the estimated annual flow rate for each plant and a cost for electricity of $0.085/kwh. Annual chemical usage was obtained from Ionics, Inc. (2000), and the unit costs for each chemical were estimated by a local chemical supplier.

Non-point source load reduction
The prescribed SOD value of 1.5 mg/L represented the impacts that non-point sources have on the river throughout the year. Therefore, to reduce the impact that non-point source loads have on the river, this SOD must be reduced. It was assumed that a percentage reduction in non-point source BOD corresponds to an equivalent reduction in the SOD.

The structural Best Management Practices (BMPs) chosen to manage the NPS loads were infiltration basins and constructed wetlands. The BOD removal rates for infiltration basins and constructed wetlands were conservatively chosen from literature-based ranges to be 50 percent and 20 percent, respectively (US EPA 1993). Infiltration basins were used in urban areas with soils dominated by sand and gravel (best represented by hydrologic soil groups A and B (Quigley 2001). Constructed wetlands were used in areas with till/bedrock soils, best represented by groups C and D (Quigley 2001). Using land use and soil type data layers from MASSGIS (2000), we found the amount of urban land of each type contributing to each reach. Using the design runoff of 1.27 cm (MA DEP 1997) and the typical percent imperviousness of urban areas of 45 percent (MA DEP 1997), the runoff volume from each soil group from an urban area into each reach was determined. By adjusting the amount of land managed by BMPs up to the total amount of urban land draining to a reach, it was possible to determine a percentage reduction in SOD from BMPs. BMP cost equations from US EPA (1999) were used. Land costs were not included. The amounts of BMPs required are also low estimates because the runoff from some storms exceeds 1.27 cm and it is expected that under climate change, the intensity of storms will increase (Meehl et al. 2000).

Costs
The costs to maintain the DO above 5 mg/l in each reach were estimated using the data above. In most cases it was found that it was less expensive to initially reduce the total phosphorus effluent concentration from each of the four wastewater treatment plants and then selectively implement BMPs along selected reaches to increase DO. Therefore the treatment plant effluent concentrations were reduced to approximately 0.16 mg/l in the Base Case and Scenario 1 and to 0.05 mg/l in all the other scenarios. Since effluent BOD levels from wastewater treatment plants are already very low, BOD point load removal was not used as a point load reduction strategy. BMPs were only implemented at the critical reaches along the river where NPS load reduction had an impact on the DO level. Because of land limitations under a few scenarios it was not possible to meet the DO target in every reach but the cumulative sum of the products of the remaining deficits and the reach lengths never exceeded 7 percent of the existing sum. Table 7.10 shows the approximate treatment costs for each scenario. Generally, the DO

management costs are not as sensitive to the scenarios as are the water quality impacts. The cost of adaptation, however, is significant since the cost of DO management raises from approximately $22.5 million in capital costs and $210 000 in annual operating costs in the Base Case to $30 to $39 million in capital costs and $300 000 to $600 000 in operating costs not including land costs in the climate change and population scenarios.

Conclusions

These results point to the need to consider the integrated impacts of temperature, streamflow, precipitation, land use and management, population, and water and wastewater management in evaluating the potential impacts of climate change upon water quality. It is found that the impacts of changes in the variables can be significant. It also raises the question of whether more attention should be placed upon these types of future changes in on-going Total Maximum Daily Load (TMDL) studies.

SEA LEVEL RISE

Introduction

The global rate of eustatic (that due to climate change) sea level rise (SLR) in the last century is in the range of 0.10 m to 0.20 m and is expected to be in the range of 0.11 m to 0.77 m over the next century (IPCC 2001a, b). In some areas of the world, SLR is also impacted by land subsidence. For example in parts of the northeastern United States, land subsidence has been 0.15 m in the last 100 years (Nucci Vine Associates, Inc. 1992). Eustatic SLR combined with land subsidence is referred to as relative SLR. The effects of SLR in the coastal zone include displacement and loss of wetlands, inundation of low-lying property, increased erosion of the shoreline, expansion of flood zones, increased water temperatures, changing water circulation patterns and more salt water intrusion into groundwater. It is also possible that due to climate change there could be changes in coastal storm patterns that alter the frequency and intensity of coastal flooding. The CLIMB study focused on periodic losses from flooding. The flooding impacts are quantified by area at-risk and cost of likely damages and management strategies under three adaptation strategies.

Using the methodology subsequently explained, our study results also indicate (Table 7.11) changes in the recurrence intervals of the 100- and 500-year events under various scenarios of SLR. Because the data used for the bootstrapping procedure encompassed only 80 years, but contained a 100-year

*Table 7.10   Capital and operation and maintenance costs to meet DO
                 targets*

Scenario number	Capital ($million)	Operation and maintenance ($million/year)
B	22.48	0.21
1	26.36	0.26
2	24.81	0.28
3	29.34	0.40
4	29.62	0.35
5	32.80	0.44
6	30.28	0.43
7	31.41	0.46
8	33.35	0.45
9	39.05	0.62
10	28.69	0.32
11	32.06	0.41
12	28.69	0.32
13	30.84	0.37

flood, the 100-year flood elevation under a scenario of no SLR occurs
approximately every 80 years in the Monte Carlo simulation. When the
Hadley Center projections of SLR are added to subsidence in the CLIMB
region for a total elevation change of 0.60 m by 2100, the recurrence intervals
decrease so that by 2100 the surge level at which damage begins, originally a
7-year flood, is being surpassed annually and the present 100-year flood is
occurring approximately on the average every four years.

Study Area

The 32 towns of the study area which abut the coast contain 110 000 hectares
of land area, and approximately 1.2 million people (US Census Bureau 2000).
There are densely populated urban areas in Boston and in the directly
surrounding cities, and rural suburbs and farmland are located along the
northern and southern edges of the study area. The diverse coastline is host to
a variety of ecosystems including sandy beaches, rocky shores, estuaries and
salt marshes.

In Metro Boston the floodplain of the current 100-year coastal storm is
approximately 11 000 hectares. The additional incremental area of the 500-
year floodplain is 1000 hectares. The entire 500-year floodplain currently
contains 1560 hectares of residential land, 260 hectares of commercial land
and 355 hectares of industrial land. It also has 23 000 residential structures.
Build out conditions are expected to be reached by 2050 when the population
of the entire region will be approximately 3.9 million people; 1.4 million will

be in the coastal towns. The employment in the commercial sector (retail and service) is expected to increase while that in the industrial sector (basic) to decline. Assuming changes in land use proportional to changes in local population and employment, by 2050, residential area in the 500-year floodplain is expected to be 1780 hectares containing 26 000 structures, commercial area will increase to 350 hectares and industrial area will decline to 290 hectares.

Storm Surge Flooding

The expected rate of SLR over the next 100 years in the study area is uncertain. While the historical rate of relative SLR in Boston has been approximately a third of a meter per century (Nucci Vine Associates, Inc. 1992), the worst-case prediction (IPCC 2001a) is an average of one-meter increase in global sea levels. Because of the uncertainty surrounding any prediction, we examined two SLR scenarios. The first, a fairly moderate estimate, assumed the regional increase predicted by the Hadley Center with the addition of local subsidence rates for a total increase of 0.60 meter over the next century. The second scenario assumed an increase in relative sea level of one meter over the next century.

The first step in this analysis was to determine for each CLIMB zone the relationship between flood levels and the area affected. Floodplain elevations were taken from the US Army Corps of Engineers (Weiner 1993), adjusting the data to 2000 levels. We used historic FEMA data for the Halloween Storm of 1991, which is the last major storm to impact the area, to estimate that property damages begin when the surge elevation reaches 2.7 meters.

This corresponds to the present 7-year surge. Since there were no feasible methods to determine the extent of the floodplain beyond the present 500-year delineation, we conservatively assumed that damages resulting from storms of greater magnitude were equal to damages caused by the 500-year storm. The present residential, commercial and industrial areas at risk in the 100- and 500-year floodplains were calculated by combining the MassGIS land use and floodplain layers. Economic damages to residential development were based on the number of units and average value per unit by census tract from 2000 US Census Bureau data. Calibrating our model to historical flood damage data from FEMA, we estimated that 40 percent of residential households get flooded in the present 100-year floodplain and 25 percent in the area between the 100- and 500-year floodplains. The flood data also indicated that flood damages to building and contents ranged from 3 to 8 percent of residential property value depending upon CLIMB zone. The damages averaged between $7000 and $18 000 depending on location. The best source of data we could find for damages to commercial and industrial property was from the Lynn Harbor study of the US Army Corps of Engineers (1990) from which we derived an estimate of $750 000 per hectare. We added 17 percent to the total damage costs as emergency costs (US Army Corps of Engineers 1990).

Table 7.11   Change in recurrence intervals for various sea level rise scenarios

Year	Zero Damage Threshold	100-Year Flood	500-Year Flood
	No change		
2000–25	4.5	100	NC
2026–50	4.7	67.6	NC
2051–75	5.1	80.6	NC
2076–2100	4.8	92.6	NC
	Subsidence		
2000–25	4.4	100	NC
2026–50	3.6	67.6	NC
2051–75	3.5	48.1	NC
2076–2100	2.6	38.5	NC
	Subsidence and Hadley Center		
2000–25	3.4	71.4	NC
2026–50	1.7	25.3	104.2
2051–75	1.1	7.1	48.1
2076–2100	1.0	3.8	13.8

Notes: (NC = not calculated)

Because our analysis was dynamic, it was necessary to account for land use changes in the floodplains over the next century. This was accomplished by assuming that all areas zoned for eventual development would be developed, and that none of the areas zoned for recreation or conservation would become residential, commercial or industrial. We assumed that no future development of wetlands would be permitted. We further assumed that development occurred in the floodplains at the same rate as in the coastal CLIMB zones. The rates of residential area change were proportional to the projected population changes and the commercial and industrial area changes were proportional to the changes in their employment.

We examined the annual impacts for the period 2000 to 2100 in light of the three possible adaptations to climate change. We examined the impacts assuming that SLR occurs gradually at a constant amount each year. We did not discount any of the future costs of property damage or flood protection; we assumed that property values appreciated at the discount rate. Because the process accounts for 100 possible scenarios, uncertainty in the timing of future storm surges is inherently included. In those scenarios that included SLR, the change in sea level for a future year was added to the sea levels corresponding to the year in the bootstrapped time-series.

A listing of the simulations is shown in Table 7.12. The Baseline analysis determines expected future damages for growth and no growth conditions considering only local subsidence rates (increasing relative sea level by 0.18 meters by 2100). Analyses were performed for both the cases of:

1. only damage from the largest event each year, assuming that it takes one year to rebuild; and
2. the possibility of damages from all three events in a year, assuming rapid reconstruction.

In a stricter version of current FEMA regulations, the Green scenario requires that all growth in the current 100- and 500-year floodplains be totally flood proofed at the time of construction and we assume that flood proofing new residential, commercial and industrial structures only nominally adds to the cost of construction. It also requires that current development be flood proofed upon sale of the structure assuming a 15-year turnover rate. The retrofitting of those structures already present in the floodplain is assumed to be 80 percent effective.

The Ride It Out adaptation assumes that buildings will be repaired to current conditions after each flood over the 100-year period with no additional flood proofing. All growth in the present 100-year floodplain is flood proofed 100 percent effectively so there are no damages to this property. This scenario most closely mimics current policy.

Under Build Your Way Out, unregulated growth is allowed in all floodplains because all current and future development is protected with retrofit or new coastal protection structures, which are all built following the second flood with a magnitude greater than or equal to the present 100-year flood. Damage is incurred until that event occurs, and as with Ride It Out, damaged structures are repaired to their previous state, allowing repetitive damages. Coastal protection in this option consists of shoreline hardening structures such as seawalls, bulkheads and revetments.

Using the bootstrapping procedure outlined above, we produced 100 possible time-series of future annual maximum sea levels for each of the scenarios in Table 7.12. Then the coastal impacts were determined for each of the 100 time-series and the resulting flooded areas and management costs and damages were averaged to obtain the results summarized in Tables 7.12 and 7.13. By comparing the Baseline runs (subsidence only) with and without growth (runs 1 through 4) in Table 7.13, it can be seen that the results of the model are not very sensitive to growth. This is because the region is already close to build out. A decrease in industrial area under growth occurs because, as stated previously, industrial employment is decreasing. Because of the results of this baseline analysis, in the other runs we no longer consider sensitivity to no growth.

The Baseline runs in Tables 7.12 and 7.13 are the flood damages expected with just subsidence if development continues as expected in the region; same as Ride It Out except without the eustatic component of SLR. Even with a small increase in SLR due to subsidence, the cumulative amounts of residential, commercial and industrial lands flooded are greater than the areas of these sectors in the floodplain. This is because many properties receive repetitive damages.

Table 7.12   *Summary of expected values of total areas flooded for 2000–2100 (hectares)*

	Model run	Residential area	Commercial area	Industrial area
1	Baseline: no growth, one event	15 714	2 334	3 030
2	Baseline: no growth, three events	20 971	3 108	4 030
3	Baseline: growth, one event	17 429	3 076	2 663
4	Baseline: growth, three events	23 350	4 121	3 526
5	Ride It Out: moderate SLR, one event	51 752	7 918	10 115
6	Build Your Way Out: moderate SLR, one event	15 834	2 786	2 525
7	Green: Moderate SLR, one event	51 597	7 785	10 192
8	Ride It Out: moderate SLR, three events	115 692	17 381	22 321
9	Build Your Way Out: moderate SLR, three events	27 803	4 902	4 331
10	Green: moderate SLR, three events	115 512	17 226	22 410
11	Ride It Out: one meter SLR, one event	90 236	14 817	18 535
12	Build Your Way Out: one meter SLR, one event	14 017	2 455	2 363
13	Green: one meter SLR, one event	89 428	5 610	18 935
14	Ride It Out: one meter SLR, three events	236 471	37 920	47 746
15	Build Your Way Out: one meter SLR, three events	26 287	4 604	4 334
16	Green: one meter SLR, three events	234 822	13 414	48 561

Table 7.13 shows that the cumulative expected damage costs to 2100 in the Baseline are $6.4 billion with flooding from one event. Comparing the Baseline to Ride It Out, Table 7.12 shows that the total flooded area more

than triples when eustatic SLR is added to subsidence. Cumulative damages increase from $6.4 billion to $20 billion. The largest residential area flooded and the highest residential damages for both the Baseline and Ride It Out runs are in Zone 4, Coastal Suburban South. Table 7.13 shows $5.9 billion in damages from the Build Your Way Out scenario with one event and moderate SLR (Run 6). After adding the Build Your Way Out adaptation costs of $3.5 billion, total Build Your Way Out cost is $9.4 billion, which is considerably less than Ride It Out. The environmental costs, however, are much greater, particularly in the lesser-developed coastal areas. In all CLIMB zones, Build Your Way Out total costs are less than Ride It Out.

Table 7.13   Costs of SLR scenarios ($million)

	Model run	Residential	Commercial/ Industrial	Emergency	Adaptation	Total
1	Baseline: no growth, one event	1 087	4 023	869	0	5 979
2	Baseline: no growth, three events	1 452	5 354	1 157	0	7 963
3	Baseline: growth, one event	1 205	4 305	937	0	6 447
4	Baseline: growth, three events	1 616	5 735	1 250	0	8 601
5	Ride It Out: moderate SLR, one event	3 563	13 525	2 905	0	19 993
6	Build Your Way Out: moderate SLR, one event	1 091	3 984	863	3 462	9 400
7	Green: moderate SLR, one event	756	2 697	587	1 766	5 806
8	Ride It Out: moderate SLR, three events	7 993	29 776	6 421	0	44 190
9	Build Your Way Out: moderate SLR, three events	1 924	6 925	1 504	3 462	13 815
10	Green: moderate SLR, three events	1 649	5 945	1 291	3 391	12 276
11	Ride It Out: one meter SLR, one event	6 131	25 014	5 295	0	36 440

(continued on next page)

Table 7.13 continued

	Model run	Residential	Commercial/ Industrial	Emergency	Adaptation	Total
14	Ride It Out: one meter SLR, one event	16 140	64 250	13 666	0	94 056
15	Build Your Way Out: one meter SLR, one event	1 820	6 703	1 449	3 462	13 434
16	Green: one meter SLR, one event	3 272	12 760	2 726	6 798	25 556

While the cumulative areas flooded under the Green Scenario are approximately the same as Ride It Out because there is no flood protection, the damages are considerably less than Ride It Out because of the flood proofing; damages decrease from $20 billion to $4.0 billion (Run 7). This requires an expenditure of $1.8 billion for flood proofing of existing residential, commercial and industrial structures, or a total damage and adaptation cost of $5.8 billion. While the area flooded under Green is greater than under Build Your Way Out, Table 7.13 shows that the damage and adaptation costs are less than Build Your Way Out so that the total cost of Green adaptation is less than both Ride It Out and Build Your Way Out for the case of moderate SLR and flooding from one annual event. This is also the case in the areal distribution of the cost impacts in the four of the five CLIMB zones. In South Coastal Urban, Zone 1 (mainly the City of Boston), the total cost differences are small between Build Your Way Out and Green; $2.3 billion versus $2.6 billion. This is because while the flood damages under Build Your Way Out are greater (due to high concentration of expensive buildings getting flooded several times before action is taken), the adaptation costs are considerably lower since it is less expensive to structurally protect a small area of the coast compared to flood proofing many individual structures in the Green approach.

The Green and Ride It Out scenarios cause less environmental damage and have lower maintenance costs than Build Your Way Out. Another drawback of Build Your Way Out is that if expected SLR magnitudes did not occur, a massive shoreline protection project with large maintenance and environmental costs would have taken place. While extra flood proofing would have occurred under the Green approach if SLR was less severe than expected, the flood proofing would provide some extra protection to floodplain users with no environmental or maintenance costs.

When climate change damage was calculated using three events or one meter of SLR, damages resulting from the Ride It Out scenario approximately doubled. If both occurred, then total damages increased more than four times compared to the initial Ride It Out scenario. Under these most extreme conditions the damage totals $94 billion over the 100-year period. In all these cases, the total costs of Ride It Out are considerably greater than either Build Your Way Out or Green. Under Build Your Way Out, area flooded approximately doubled with three annual flood events and moderate SLR compared to one annual flood event. There, however, were slightly less flooded area and damages with one meter of SLR than with moderate SLR because the protection is built sooner under the one meter SLR scenario. Under the worst case of severe SLR and three flood events, the total damages and adaptation cost of Build Your Way Out (Run 15) increase to $13.4 billion compared to the situation of moderate SLR and one annual event of $9.4 billion. Green damages and total adaptation costs also increase under separate and combined increases in annual flood events and SLR. Under the worst case of severe SLR and three flood events, the total damages and adaptation cost of Green increase to $25.5 billion compared to the situation of moderate SLR and one annual event of $5.8 billion.

An analysis of costs by CLIMB zone (not shown) found that in the scenario of three annual damaging events and moderate SLR, the total costs of Build Your Way Out and Green are similar in all zones except Zone 4 (South Coastal Suburban) where the cost of Build Your Way Out is approximately $1.4 billion higher. Here the cost of structurally protecting a long coastline that is not densely developed is not cost effective compared to flood proofing. In the case of one meter of SLR and either one or three annual damaging events, the major difference between Build Your Way Out and Green is the decrease in total costs of Zone 1 (mainly the City of Boston) under Build Your Way Out. Flood damages have decreased here because the second 100-year flood occurs faster than under the moderate SLR. Adaptation costs are also less in this region because a large amount of expensive property can be protected with a relatively short length of coastal protection. In the most extreme case of one meter of SLR and three annual damaging events in all zones the total costs of Build Your Way Out are less than the total costs of Green adaptation.

Conclusions

Our findings on adaptation to increased storm surge impacts support those of others; it may be advantageous to use expensive structural protection in areas that are highly developed and take a less structural approach in less developed areas and/or environmentally sensitive areas (Titus et al. 1992; Darwin and Tol 2001; Yohe 2000; Neumann et al. 2000). Besides being more cost effective, the less structural approaches are no-regrets or co-benefit policies, are environmentally benign, and allow more flexibility to respond to future uncertain changes. While uncertainty in the expected rate of SLR and damages makes planning difficult, the results also show that no matter what the climate change scenario or the location, not taking action is the worst response.

RIVER FLOODING

Introduction

Estimates of flood-related impacts resulting from the different climate, socioeconomic and adaptation scenarios are based, in part, on flood insurance studies developed by FEMA's National Flood Insurance Program (FEMA 2002). These studies involved the modeling of river flooding using detailed topographic data and standard engineering techniques in order to produce Flood Insurance Rate Maps (FIRMs), which assess flood hazards. Additionally, we utilize a digital data layer of the approximately 2300 sub-basins in the CLIMB study region, as defined by the USGS Water Resources Division (MassGIS) and FEMA's Flood Insurance Claims Database.

For present conditions, we associate the 24-hour, 100-year storm with the 100-year floodplain determined in FIRMs. This is based on the fact that the time of concentration of watersheds in the area is less than 24 hours (time of concentration is a hydrologic term that refers to the time of travel of water in a watershed). Since the FIRM information is entirely expressed in relation to events with recurrence intervals of 100 and 500 years, we adopt the runoff associated with the current 100-year floodplain as a reference value. For any given watershed, runoff is normalized using that reference value. This allows us to consider only the relative magnitude of runoff, which is likely to be homogenous throughout a watershed. In contrast, absolute flows (measured in cubic feet per second) would demand values for a multiplicity of transects along each stream, adding spatial complication to the model.

Using GIS, we can compute for any given town the total areas flooded by 100- or 500-year runoff. In order to determine what property is affected by flooding, we concentrate on residential, industrial and commercial land uses. Using the build out analysis information we also calculate the area that would get flooded for each land use type if all developable land were to become developed.

Adaptation scenarios are incorporated in this dynamic model in two forms – by changing the relationship between average flood-related expenditures and average property values, and through future land use scenarios. The Ride It Out scenario assumes that structures are repaired to their pre-disaster condition. Therefore, since the intensity of events will increase under climate change, the ratio of expenditures to value will tend to rise with time. On the other hand, under the Green and Build Your Way Out scenarios, adaptive measures such as flood proofing are implemented in early stages in order to prevent future damages. The difference between the two scenarios is that in Build Your Way Out only properties that have been flooded once are flood proofed, while in Green it is assumed that flood proofing is completed for all vulnerable properties within the first 15 years. Also, one of the policy alternatives available to cope with climate change impacts in the Green scenario is to establish regulations and incentives to prevent the construction of new infrastructure in flood-prone areas.

Simulation

The CLIMB river flooding model was used to simulate total river flood damages over the period from 2001 to 2100 under a Baseline scenario and under scenarios of climate change. In the Baseline scenario, the frequency and intensity of rainfall events is assumed to be the same throughout the 21st century as it was over the second half of the 20th century. Since flooding is an inherently stochastic phenomenon, the results shown below are based on averages of the results of multiple bootstraps and repeated stochastic simulations.

In order to represent the increased intensity of storm events under climate change, the entire exercise is repeated with a built-in adjustment whereby the assigned rainfall amounts are increased by 0.31 percent each year from 2001 to 2100. This generates an expected increase over the course of the century that is consistent with projections from the Canadian Climate Center (Kharin and Zwiers 2000). Again, repeated simulations are conducted and the average values of damage indicators are recorded. This entire procedure – the Baseline simulations plus the climate change simulations – is repeated for each of the three adaptation scenarios: Ride It Out, Build Your Way Out and Green.

Findings

Table 7.14 presents information for the accumulated value of damages to commercial, industrial and residential properties. If the same property is damaged twice it is counted twice. The same property can be damaged as many as three times in one year, although this seldom occurs. (These are damage costs only and do not include adaptation costs.) Note that in the worst case, incremental property damages due to climate change are in the order of $30 billion. However, this number can be reduced by about two-thirds under the Green adaptation strategy.

*Table 7.14   Dollar value of floods under baseline and climate change
             scenarios*

	Total damage costs 2000–2100, $millions (three events)			
	Commercial	Industrial	Residential	Total
Ride it out				
No climate change	6 226	22 741	1 789	31 155
Climate change	12 121	41 096	3 964	57 182
Increment	5 895	18 355	2 175	26 027
Build Your Way Out				
No climate change	5 834	19 724	1 629	27 187
Climate change	12 573	42 301	4 056	58 930
Increment	6 739	22 577	2 427	31 743
Green				
No climate change	2 903	6 832	693	10 428
Climate change	5 322	13 097	1 126	19 546
Increment	2 419	6 265	433	9 118

SURFACE TRANSPORTATION

Introduction

The principal way in which climate change will affect the transportation
system is through extreme climate events. In particular, events that produce
significant flooding or snowfall will have a negative impact on the
transportation system. The main impacts of SLR will also be felt during
extreme events when storm tides will be higher, increasing the frequency and
severity of coastal flooding. This section addresses potential impacts of such
climate change on the surface transportation infrastructure system, with an
emphasis on road transportation.

In economic terms, the impacts of extreme weather events on the
transportation system are of two types. The first is the damage inflicted upon
infrastructure, such as flood damage to road, rail and bridges. The second is
the economic cost of interruptions in the operation of the transportation
systems, which prevent employees from going to work, shoppers from getting
to stores, goods from being delivered to factories, warehouses and stores. Our
modeling efforts address the second type. We are asking the following
question: what are the costs of extreme weather events defined in terms of
disruptions to the transportation system? This leads to a second question: if
extreme weather events increase in the future due to climate change, how will
these costs increase?

Approach

The key research asset is a model that is capable of simulating the flows of road and rail traffic (including private and public road transportation) in the metropolitan area under a variety of conditions. This model is first run under normal circumstances to provide Baseline values for the volume of travel and the amount of time spent in travel. Then a scenario is designed to identify those areas that are flooded (so that no trips begin and end there) and those network links that become impassable. The model is rerun and the results are compared to the initial run to determine how many lost trips and how much extra travel time may be attributed to the weather event. This provides a basis for estimating the transportation related costs of more frequent and more extreme weather events under various climate change scenarios.

The Central Transportation Planning Staff (CTPS 2000), which serves the Metropolitan Planning Organization (MPO) for the Boston area developed a highly sophisticated Urban Transportation Modeling System (UTMS) that is capable of producing spatially detailed projections of transportation activities. This model was run as described above – with and without the effect of the extreme weather event – for a number of current and future scenarios assuming weather events of different type and magnitude. The UTMS breaks traffic generation into four stages: trip generation, trip distribution, modal split and traffic assignment (Figure 7.4).

In the trip generation stage, the number of trips originating in each zone is estimated on the basis of zone characteristics. Once a balanced set of trip origins and destinations are generated, zone-to-zone flows are estimated in the trip distribution stage. The zone-to-zone flows are then separated in the modal split stage into trips by drive-alone car, carpool, public transit, public–private combinations, bicycle and so on. This is achieved using a model that relates the probability that an individual will choose a particular mode to characteristics of the available modes relative to one another, and to characteristics of the individual such as income, occupation and gender. (Train 1986). This effectively translates the zone-to-zone person trips into different kinds of zone-to-zone vehicle trips. The traffic assignment stage simultaneously assigns all trips to specific zone-to-zone paths made up of links on the road network. This is done by means of a model that takes account of the fact that the speed of travel along each link decreases (due to congestion) with the amount of traffic assigned to it. An equilibrium assignment is determined whereby no driver could reduce his or her total travel time by switching to another path (Sheffi 1985.) The end product is the flow of traffic and the average speed of travel for each link. Aggregate vehicle miles can be calculated by summing up the product of link flows and link lengths, and an aggregate average speed can be taken as a weighted average of link speeds.

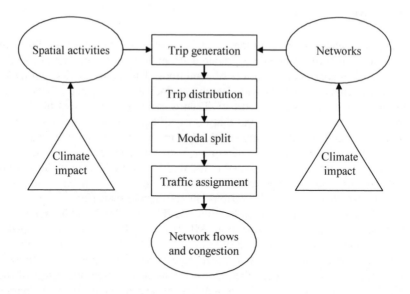

Figure 7.4   The Urban Transportation Modeling System (UTMS)

Simulating the Effect of Flooding Using UTMS

In order to simulate the effect of floods we identify first those areas that are inundated so that we would expect trips to or from them to be canceled, and second those network links that are inundated and therefore impassable. Creation of the flood scenario information involves overlaying the flood map with maps showing the 986 traffic analysis zones and the more than 22 000 network links from the UTMS. Figure 7.5 shows such an overlay. Identification of overlaps in 100- and 500-year flood zones with a detailed representation of the road network then yields information of those portions of the transport system that are susceptible to climate impacts.

Scenarios

In order to capture a sufficient range of impacts to provide a basis for predicting the impact of climate change, models were run under a number of different scenarios shown in Table 7.15. The model was run for two years, 2003 and 2025. The 2003 runs reflect contemporary patterns of travel and the current state of the road network. The 2025 runs, which are the most farthest into the future of the CTPS projects, was considered to represent patterns of travel given anticipated growth in the Boston metropolitan area and the road network with anticipated extensions.

Table 17.16 shows results for 2003. The first column summarizes the base (no flood) scenario. All other numbers in the table represent deviations from these values under the various flooding scenarios. So, for example, there are 14 619 585 trips taken each day in the base scenario and this number is reduced by 88 914 in the riverine flooding, 100-year flood scenario. The number of trips is always expected to decline because of a flood, but the effect of the flood on the other two indicators, vehicle miles traveled (VMT) and vehicle hours traveled (VHT), is ambiguous. In the case of VMT, the value may go down because there are fewer trips. However it may go up because the trips that are made in some cases take more circuitous routes than in the base scenario due to the flooded links. Similarly, VHT may go down due to fewer trips or up due to greater congestion. The increased congestion is reflected in the overall reduction if average speed in the flood scenarios. Similar procedures were followed to determine conditions in 2025.

Figure 7.5   *Overlay of flood map with traffic analysis zones (Dark gray indicates the 100-year floodplain; light gray indicates the 500-year floodplain.)*

Table 7.15   Scenarios for CLIMB transportation analysis

	No flooding	River flooding
Current (2003)	Baseline conditions given current socioeconomic data and transport infrastructure	Inland flooding under current socioeconomic data and infrastructure
Future (2025)	Baseline condition given projected socioeconomic data and infrastructure	Inland flooding under projected socioeconomic data and infrastructure

Table 7.16   Results of 2003 flood simulations

	100-year flood event	
	Base	Riverine flooding
Links deleted	0	445
Total trips	14 619 585	-88 914
VMT	135 916 944	2 421 776
VHT	3 862 297	181 930
Avg. speed	35.2	-1.0
	500-year flood event	
	Base	Riverine flooding
Links deleted	0	673
Total trips	14 619 585	-150 789
VMT	135 916 944	4 183 432
VHT	3 862 297	307 459
Avg. speed	35.2	-1.6

Notes: VMT Vehicle miles traveled; VHT Vehicle hours traveled

Assessing Likely Effects of Climate Change

We then make an assumption of the rate of increase in the magnitude of rainfall events and re-estimate delay and lost trips. The difference between the two sets of estimates may be attributed to climate change. Note that this analysis is limited to the effect of climate change on river flooding events, but we believe that the relative magnitude of the increase in delays and lost trips due to coastal flooding events will be comparable. The analysis proceeded as follows:

1. We obtained a table of the three largest warm weather precipitation events in Boston for every year from 1950 to 1999.
2. As a projection of rainfall events in the future we randomly assigned one year from 1950 to 1999 to each year from 2000 to 2100. We assumed that

the rainfall events that occurred in the former year will also occur in the latter year.

3. We extrapolated the values of delay and lost trips between the 100-year river flood, which occurs with a 24 hour rainfall of 6.75 inches and the 500-year river flood that occurs with a rainfall of 8.22 inches. This allowed us to estimate delay and lost trips for a rainfall event of any size. Based on the assumption that the 10-year flood, which occurs with a rainfall of 4.5 inches, is a 'no damage' event, we set the values of delay and lost trips to zero for every event smaller than 4.5 inches.

4. Values for delay and lost trips were summed over the 100 year period.

5. To represent the effects of climate change, we assumed that the magnitude of extreme rainfall events would increase at a rate of 0.31 percent per year throughout the 21st century (Kharin and Zwiers 2000).

6. We repeated steps 1–4 under this assumption. This allows us to compare the total delay and lost trips with and without climate change.

Because of the probabilistic nature of the random assignment, this entire exercise was repeated 100 times and final results are averages over those estimates.

The results in Table 7.17 indicate a reasonable expectation that the magnitude of hours and trips lost as a result of extreme rainfall events in the metropolitan area will be much higher under a scenario of climate change. The table indicates that under a climate change scenario, aggregate traffic delay due to flood events over the course of the 21st century may be expected to increase by about 80 percent, and lost trips over the same period may be expected to increase by 82 percent, over delay and lost trips that would be expected in the absence of climate change. We should be careful, however, not to overstate the significance of this result, since the aggregate numbers are still relatively modest relative to the total number of trips and travel hours in the system shown in Table 7.16. Rather we might say that a problem that could be described as a significant nuisance will become noticeably and increasingly worse as the climate changes over the course of the century.

Table 7.17   Aggregate delay and lost trips 2000–2100

	No climate change	Climate change
Delay (hours)	1 196 634	2 153 948
Lost trips	516 130	940 707

ENERGY DEMAND

Background

To identify climate impacts on the energy sector, we use a degree-day methodology to estimate energy demand under various climate scenarios. Degree-days are a common energy accounting practice for forecasting energy demand as a function of heating degree-days (HDD) and cooling degree-days (CDD). The degree-day methodology presumes a V-shaped temperature–energy consumption relationship as shown in Figure 7.6 (Jager 1983). At the balance point temperature (the bottom of the V-shaped temperature-energy consumption function) energy demand is at a minimum since outside climatic conditions produce the desired indoor temperature. The amount of energy demanded at the balance point temperature is the non-temperature sensitive energy load. As outdoor temperatures deviate above or below the balance point temperature, energy demand increases proportionally. Energy demanded in excess of the level at the balance point temperature is the temperature-sensitive energy load.

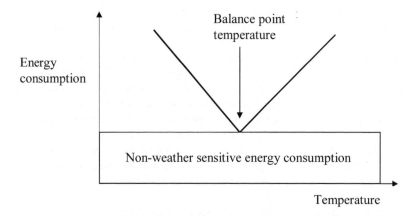

Figure 7.6   Theoretical relationship between temperature and energy use

The actual balance point temperature of an energy system varies depending on the place-specific characteristics of the building stock, non-temperature weather conditions (for example humidity, precipitation and wind) and cultural preferences (Nall and Arens 1979; de Dear and Brager 2001). We use a quantitative approach to objectively tailor the balance point to the attributes of the Massachusetts energy infrastructure such that the functional relationship is optimally specified. Similar to the methodology used by Belzer and colleagues (Belzer et al. 1996), we iteratively regressed statistical time-series models with base temperatures at 5°F intervals choosing the base temperature producing the highest R-square as the balance point temperature. In this way the temperature explaining the largest share of changes in energy use is objectively designated as the balance point temperature. We find a balance point temperature for electricity of 60°F in the residential sector and 55°F for commercial sector. The balance point temperature for fuels used for heating is 65°F in the residential sector and 60°F in the commercial sector.

Methodology

Our methodology is a two-step modeling and estimation procedure. First we use monthly time-series data to quantify the historic sensitivity of end use energy demand to temperature variables while controlling for socioeconomic factors such as population size and energy prices as well as daylight hours.

We independently estimate Massachusetts' residential and commercial energy demand sensitivities to temperature variables because potentially different energy use – temperature relations exist between economic sectors (Sailor and Munoz 1997; Sailor 2001). Industrial energy demand is not estimated since previous investigations (Elkhafif 1996; Sailor and Munoz 1997) and our own findings indicate that it is non-temperature sensitive.

Furthermore, for each sector the demand for electricity and heating fuels is separately estimated. We assume the separation of energy forms used predominantly for heating (for example natural gas and fuel oil) and those for cooling (for example electricity) is important because climate change is anticipated to have unique impacts on the use of each form of energy and, subsequently, on the different energy delivery systems. Analyses focusing on total energy use may find only negligible changes in energy use or expenditure given the potential for changes in cooling and heating energy to offset one another. While the implications may not be significant in physical energy terms, the implications may be significant in terms of the large capital costs associated with cooling energy system expansion and heating energy system contraction.

In the second part of the analysis we estimate future energy consumption under various climate change scenarios by employing the energy sensitivity relationships developed in the first step of our analysis. While this study focuses on the temperature effects of climate change, other weather variables not modeled here – such as humidity, precipitation and wind – may have a significant impact on energy demand if they change as climate changes.

Data Sources

Energy data is from the US Energy Information Administration (US EIA). Monthly electricity sales and price to residential and commercial end users are from *Electric Power Monthly* (US EIA various years). Monthly natural gas sales to residential and commercial end users are from the *Natural Gas Monthly* (US EIA various years). Monthly sales of heating oil (distillate fuel oil No. 2) to all end users are published in *Petroleum Marketing Monthly* (US EIA various years). Because sales to end use sector data are not available and the majority of heating oil is consumed by the residential sector, we assume that all heating fuel sales are to residential end users. The heating oil sales and price data cover the January 1983 to December 2001 period. All prices are adjusted for inflation using the Bureau of Labor Statistics' consumer price indices for electricity and fuels, respectively, in the New England region (BLS 2003).

Population and employment data for the commercial sector are consistent with those described above. Climate data were generated with the bootstrapping approach introduced above. The latter is used to calculate monthly HDD and CDD. We further use the hours of daylight on the 15th day of each month, calculated as the time elapsing between sunset and sunrise, as a proxy for the number of daylight hours per month in Boston (NOAA 2003). Those hours are used in regression analyses alongside the HDD, CDD and price variables to derive fuel specific energy demand in the region (for a detailed discussion of the statistical approach and results see Amato et al. 2005).

Projections of Energy Use Under Future Climate Scenarios

To generate scenarios of future energy demand by energy type and end use sector, we use the time-series analysis in conjunction with various climate change scenarios to project future energy use. Price parameters are held constant at 2000 levels since future prices are unknown. Not all of the future changes in the region's energy consumption are due to climate change; part is driven by trends in proliferation and efficiencies of energy using technologies, changes in household structure and size, as well as changes in income in the region. We assume these factors continue to change as they have over the analysis period.

The projections for monthly energy use by the residential sector under the assumption of climate trends from the Canadian Climate Center and from the Hadley model are shown in Figures 7.7–7.12. These results are averages of model results from 100 bootstrapped climate series. To the right of each figure is a bar chart of the percent change in monthly energy use attributable to climate change – the difference between the amount projected for a year under a climate change scenario and the amount for the same year under a non-climate change scenario. Results for the commercial sector are qualitatively similar to those of the residential sector (see Amato et al. 2005 for details).

Results indicate that residential and commercial energy demand in Massachusetts are sensitive to temperature and that a range of scenarios of climate change may noticeably decrease winter heating fuel and electricity demands and increase summer electricity demands. These findings suggest a need to incorporate the impacts of climate change into regional energy system expansion plans to ensure adequate supply of energy both throughout the year and for periods of peak demand.

The energy demand responses modeled here provide estimates of the impacts of climate change under a business-as-usual scenario and assist in identifying the need for anticipatory adaptation. Anticipatory adaptation could alter a region's energy demand response function to more effectively correspond with future climatic conditions via planned adjustments in the attributes of temperature-sensitive infrastructure and energy technologies such as building thermal shells, air conditioners and furnaces.

Identifying potential impacts for the region now is important because the energy industry is extremely capital intensive and as a consequence the flexibility of policy induced changes in energy generation and demand trajectories over the short and medium run is limited (Grubler 1990). In the long run, as the capital stock naturally turns over, building codes may be changed to calibrate the thermal attributes of the building stock to expected future climates (Camilleri et al. 2001). However, such changes need to be implemented in the relatively near term or the building stock will become increasingly maladapted to climate. In the near term, polices such as urban shade tree planting and installation of high albedo roofs can begin to modify the thermal characteristics of the Massachusetts energy infrastructure in order to reduce space-conditioning energy use.

PUBLIC HEALTH

Introduction

Differential levels of certainty exist for the relationship between human disease outcomes and meteorological factors, which will be affected by global climate change. Sufficiently reliable information on the many potential ramifications of climate change for the spread of diseases, and thus regional morbidity and mortality, is scarce. In contrast, impacts of changes in temperature can more easily be discerned from the historic record and based on sufficiently sound empirical grounds to extrapolate into the future.

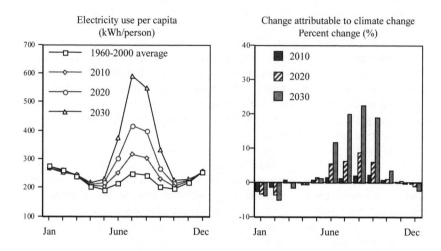

Figure 7.7   Residential electricity per capita under Canadian Climate Center climate scenarios (results are averages across 100 bootstrapped scenarios)

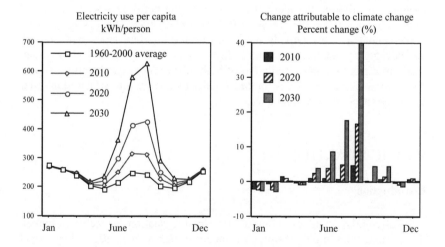

Figure 7.8   Residential electricity per capita under Hadley climate scenarios (results are averages across 100 bootstrapped scenarios)

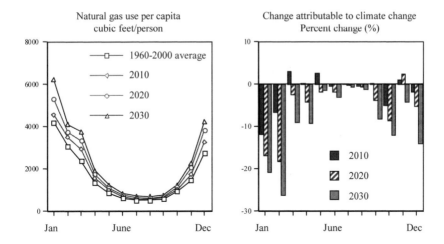

*Figure 7.9   Residential natural gas per capita under Canadian Climate
            Center climate scenarios (results are averages across 100
            bootstrapped scenarios)*

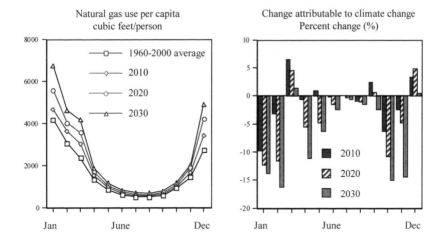

*Figure 7.10   Residential natural gas per capita under Hadley climate
            scenarios (results are averages across 100 bootstrapped
            scenarios)*

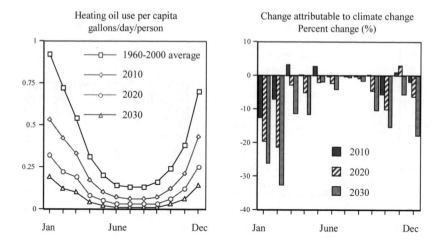

*Figure 7.11   Residential heating oil per capita under Canadian Climate
           Center climate scenarios (results are averages across 100
           bootstrapped scenarios)*

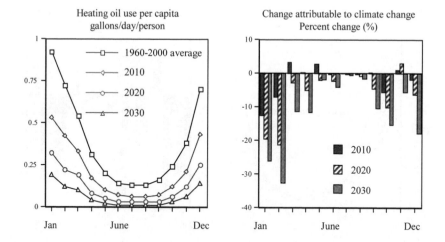

*Figure 7.12   Residential heating oil capita under Hadley climate scenarios
           (results are averages across 100 bootstrapped climate
           scenarios)*

Consequently, the CLIMB project focuses on temperature-related (as a subset of weather-related events) mortality.

Exposure to temperature extremes, such as those experienced during heatwaves and cold spells, is associated with rapid increases in mortality (Huynen et al. 2001). The combination of sudden increases in mortality during extreme temperature events and the downward sloping non-extreme temperature–mortality relation produces an overall temperature–mortality relation as schematically shown in Figure 7.13.

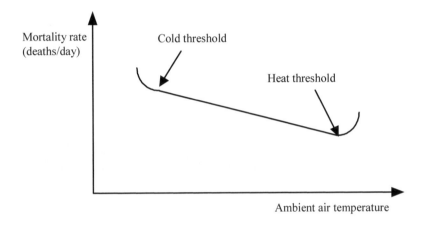

Figure 7.13   Theoretical temperature–mortality relationship

The hot and cold temperature extremes beyond which mortality rates deviate upward from the generally linear temperature–mortality relation are a population's threshold levels. We refer to the high and low temperatures where mortality rates sharply increase as the population's heat threshold and cold threshold, respectively.

Estimates of the temperature-related mortality in Metropolitan Boston are derived under the socioeconomic and climate change scenarios discussed above. The Massachusetts Department of Public Health provided the mortality records used in this study. The records were presorted on the basis of the deceased residing in one of the 101 towns in the CLIMB region. We retain mortality observations on the basis of residence because we are interested in examining changes in mortality attributable to Metropolitan Boston's weather. Accordingly, mortality records were excluded for individuals who died in the region but did not reside within the region.

Consistent data records extended for the years 1970 to 1998 and contained mortality information for 754 778 deaths, or on average 25 000 deaths per year, in the CLIMB region. The mortality data are aggregated to daily totals for each day between 1 January 1970 and 31 December 1998 generating 10 592 observations in the data set. The daily mortality rate – expressed as deaths per million people – is the daily mortality divided by the population in the CLIMB region in the respective year (see Figure 7.14).

The mean daily mortality rate during the 1970 to 1998 period is 23.5 deaths per million people with a standard deviation of 3.7. The large spike in the mortality rate on 3 August 1975 to 66 deaths per million people occurred a day after Boston experienced an August all-time maximum temperature of 102°F. On the day of the extreme heat event the mortality rate increased by 100 percent followed by a 200 percent increase on the following day, relative to the 1970 to 1998 average August daily mortality rate.

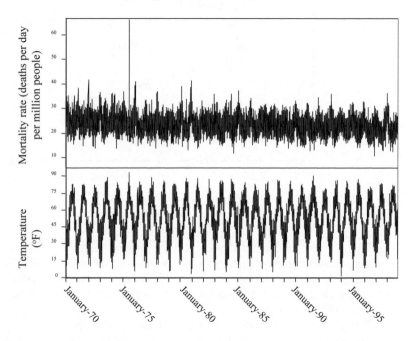

Figure 7.14   CLIMB region daily mortality rate and average temperature, 1970–1998

Methodology

Our methodology for estimating climate-induced changes in mortality for the CLIMB region entails a three-step approach. We first determine heat and cold threshold temperatures indigenous to the region. We then statistically analyze changes in mortality associated with daily temperatures in excess of the thresholds derived in the first step. Finally, in the last step we use the threshold-mortality relationships along with population and climate forecasts to simulate potential future mortality impacts in the CLIMB region.

We control for the seasonal fluctuations in mortality rates by capturing monthly mortality variation using a fixed-effects regression model. This approach enables us to compare the actual daily mortality rate to the expected mortality rate within a given month and, consequently, observe if mortality anomalies at extreme temperatures are present.

Maximum daily temperature and incidence of snowfall are the independent weather variables in the statistical model. The maximum temperature is used rather than an average or minimum temperature due to the fact that the region has a colder climate than the US average and, hence, the population has acclimatized to cooler weather but remains unaccustomed to heat extremes. The rational is similar to that of Yan who uses minimum temperatures in assessing mortality rates in relatively tropical Hong Kong where the population is unaccustomed to cold extremes (Yan 2000). We also controlled for the incidence of snowfall given some evidence in the literature of a strong association between snowfall and increases in mortality (Rogot and Padgett 1976; Baker-Blocker 1982; Gorjanc et al. 1999).

Results

Temperature thresholds
The results of our quantitative investigation show that the variables for all months (February–December) are strongly associated with changes in the mortality rate and show a statistical difference from the 27.2 deaths per million people in January. August has the largest decrease in mortality rate at 4.4 compared with the January rate. An annual trend variable indicates that the region's mortality rate has been decreasing at an average annual rate of 0.1 deaths per day per million people over the period under investigation. With the annual trend variable we assume a uniform trend across all monthly mortality rates. This assumption is supported by a variety of preliminary analyses and hypotheses tests.

Days in which the maximum temperature reaches between 80°F and 89°F have no significant association with changes in the mortality rates on the current day as indicated by their respective heat threshold dummy variables. However, each of the dummy variables for maximum temperatures of 90°F or higher suggest a significant increase compared to the expected mortality rate. The results unambiguously suggest a heat threshold of 90°F for the CLIMB region, which coincides with the temperature at which heatstroke, heat cramps, and heat exhaustion begin to occur (NWS 2002) and is consistent with previous findings for Boston (Kalkstein and Davis 1989).

The results of the model indicate no clear association between extreme cold and mortality rates after controlling for season. Therefore, we elect to use 28°F as the cold threshold level for the CLIMB region, which is the coinciding maximum temperature threshold below which temperatures occur at the same frequency as temperatures 90°F or above. Our underlying assumption is that the population adapts to a particular temperature probability distribution and is susceptible to temperatures occurring at the tails of the distribution.

Temperature–mortality relationships

On the basis of the analysis presented above, we designate 90°F as the heat threshold and 28°F as the cold threshold temperatures. We include in the model a trend variable of the heat-related mortality to capture changes in the sensitivity of mortality to extreme heat over the period of analysis because some recent research suggests a de-sensitizing of populations to extreme heat events (Davis et al. 2002). Additionally, we extend the lag lengths on the cold threshold to five days in light of the prolonged effect on mortality rates. Statistical results indicate that daily mortality rates in February to December are lower than in January, with the largest difference appearing in August. The annual trend variable indicates that over the 1970 to 1998 period the daily mortality rate had an annual decrease of 0.10 deaths per day per million people.

Extreme heat has an immediate and prolonged effect on mortality rates. For example, our results suggest that on an August day in 1970 with a temperature of at least 90°F the mortality rate increases by 2.21 deaths per day per million people, or by 9.8 percent, compared to a day with temperature in the reference range. The sensitivity of mortality rates to extreme heat events has, however, decreased over the period of analysis by, on average, 0.05 per year. Consequently, the results suggest that heat-related mortality decreased by roughly 50 percent over the 1970 to 1992 period. The results also indicate that the effect of extreme heat on mortality is prolonged as the day after an extreme heat event has a mortality rate 2.37 higher and the following day 0.76 higher relative to a non-extreme temperature day. Similar to the immediate effects of extreme heat on mortality the delayed effects have also been decreasing by approximately the same amount over the period of analysis.

The effect of extreme cold on mortality is inconclusive. For example, mortality rates are 0.62 lower on the current day if the maximum temperature is below 28°F and then, three and four days later, increase by 0.59 and 0.42, respectively. The delayed effect of extreme cold on mortality has also been found in other studies (Donaldson and Keatinge 2002).

Snow on the current and previous day is positively and significantly associated with mortality rates. The incidence of snowfall is associated with increases of 0.69 on the current day and 0.63 the succeeding day. The results suggest that mortality rates are more sensitive to snowfall than to days with maximum temperatures 28°F or below. Such findings may, in part, explain the 22 percent increase in deaths from ischemic heart disease in Boston in the week following a blizzard (Glass and Zack 1979). Similar high mortality sensitivity to snow has been observed in Minneapolis–St. Paul, Minnesota (Baker-Blocker 1982).

Since our choice of a 28°F cold threshold was not strongly supported by statistical criteria and the temperature–mortality relationship was similarly ambiguous for the effects of temperature on cold-related mortality, we choose not to investigate further in our scenario analysis potential impacts of future climates on cold-related mortality. Suffice to say, though, the inconclusive results of the historical analysis are by themselves noteworthy and correspond to similar findings in the literature (Kalkstein and Greene 1997).

Climate change scenarios
We use scenarios of future weather specifically developed for the region to assess changes in annual heat-related mortality, which is the annual sum of the daily mortality predicted by the statistical model. The simulation results show a decline in regional mortality. That decline is more pronounced under the two climate model scenarios in which the frequency of extreme heat events increases. These findings are predominantly caused by the assumed continuation of past trends in mortality declines and are discussed in more detail in the following section. The simulation results further show on average slightly higher mortality until about 2010 under climate change, compared to the base case. Temporarily, the number of heat-related deaths is positive. From 2010 onward, mortality declines more rapidly under climate change than without it and from approximately 2012 onward, the number of deaths actually declines as the number of heat events increases. One explanation behind this observed reversal lies in the effects that repeated events may have on a population's adaptive behavior – the more frequent the number of events, the more prepared the population is to deal with them.

Conclusions

Our analysis indicates that there is no clear, discernible cold threshold for the population in the Metropolitan Boston region, nor is there a strong statistical association between historically observed low temperatures and cold-related mortality. In contrast, we do find a heat threshold of 90°F and a strong influence on heat-related mortality by the number of days above that threshold. Using the derived statistical relationship between mortality rates and the number of days above 90°F in conjunction with simulations of alternative future climate, we find that more frequent occurrence of heat episodes will not result in higher mortality in Metropolitan Boston, presumably because of more aggressive adaptation to those events.

These findings are subject to several interrelated assumptions about socioeconomic characteristics in the region. First, many key characteristics of the regional population have not been explicitly considered here, such as its age distribution, ethnic mix or economic prosperity, many of which may influence the population's susceptibility to weather and climate, and many of which are likely to change over the simulated one hundred years.

Second, a variety of behavioral and technological changes may likely occur as the region's climate changes. For example, an increase in the use of air conditioning (AC) will likely reduce susceptibility to heatwaves for those individuals who have access to air conditioned space. Similarly, improvements in health care, use of early warning systems for individuals most prone to changes in temperature and its often associated low outdoor air quality will likely reduce respiratory and cardiovascular stress, and thus also likely reduce heat-related morbidity and mortality. Other adaptations to extreme temperatures include changing the extent to which individuals remain outdoors. Already, people in climates with extreme heat or cold periods have found ways to reduce exposure by moving from one cooled or heated space to another (for example from the home to the car to the store and back) with little time spent outside.

Increases in access to air conditioned space, many improvements in health care and a multitude of behavioral changes have been observed in the past two decades on which the statistical analysis of this chapter is based. These improvements largely determine the sign and magnitude of the trend variable discussed above. For that variable to continue its relevance over the simulated future one hundred years requires that the factors contributing to it remain, in aggregate, comparable to the past. However, rates of expansion of air conditioning, for example, may likely be lower in the future than they have been in the past because continued proliferation tends to be more difficult as near 100 percent saturation is reached. Also, expansion of air conditioning itself is not without problems as it increases regional energy consumption, contributes to urban heat island effects, and potentially exacerbates health risks associated with low outdoor air quality.

Improvements in health care have likewise been quite significant over the last two decades. Not only are there now regular weather and health warning systems in place – from forecasts in daily news media of heat indices and chill factors to pollen counts and ground level ozone concentrations – but also the population and its health care system have found numerous ways to deal with cold spells or heatwaves. Whether a continued high rate of improvements in warning and health care systems can possibly be maintained over the next one hundred years is open to debate.

As any of the factors that influence the trend variable reduces its impact on lowering cold- or heat-related mortality rates, the other factors need to make up for it, so that the results continue to hold. Alternatively, additional adaptations to climate change may be needed. For example, the region has seen only few efforts to increase the use of shade trees to decrease albedo, increase moisture retention and thus contribute to local cooling. Similarly, little new construction uses materials or designs that reduce a building's albedo, its heating and cooling needs, and thus energy consumption and impacts on local air quality. Such engineering approaches to prepare the local building stock to a changing climate, together with appropriate zoning and transportation planning could go a long way in reducing, for example, urban heat island effects, which may be exacerbated by climate change.

The results presented above suggest that future reductions in heat-related mortality are likely under a wide range of climate scenarios. For these results to be achievable requires aggressive investments in all areas ranging from health care to space cooling to smart land use, as well as potentially drastic behavioral adjustments of the local population. On the one hand, such adjustments will need to be large, yet given past experience seem doable. On the other hand, they will quite likely entail major changes in lifestyles in the region. The analysis presented above calls for public debate on these trade-offs and necessary investments in climate change mitigation and adaptation strategies.

INTEGRATION OF IMPACTS AND ADAPTATION OPTIONS

Impacts

The emphasis of the CLIMB project was on the integration of climate and demographic changes on infrastructure in Metro Boston and on examining these impacts with a common framework. Based upon the results of this research, it is also possible to examine how impacts in one sector will impact another sector. Table 7.18 shows this for the Ride It Out scenario. Impacts upon the environment and economic and societal sectors are also shown. Ride It Out impacts are summarized in boldface in the table. Reading the table horizontally shows the impacts of one sector upon another. Reading the table vertically shows possible impacts upon that sector from all the systems we analyzed.

As can be seen, the Ride It Out negative impacts of one infrastructure system in most cases will also negatively impact the performances of other infrastructure systems. River flooding most negatively impacts the other sec-

Table 7.18   Integration of infrastructure impacts

	Energy	Health	Transport	River flooding	Sea level rise
Energy	**Summer: more electricity demand, brown outs and local emissions. Winter: less gas and heating oil demand.**	Summer: decrease in air quality; higher morbidity. Winter: less heat-related mortality.	Summer: Loss of rail service, traffic signals and draw bridges; disruption of air traffic.	NA	NA
Health	NA	**Summer: slightly higher mortality until about 2010; increased emissions-related illness.**	NA	NA	NA
Transport	Increased energy demand.	Reduced public safety.	**Increased travel time and miles; loss of trips.**	NA	NA
River flooding	Possible disruption in local deliveries.	Increased pathogens in water supply.	(See Transport)	**Temporary loss of land and land activity.**	Increased flooding impacts.
Sea level rise	NA	NA	(See Transport)	Could increase river flood losses.	**Permanent loss of some coastal land; temporary loss of land and land activities.**
Water supply	Possible loss of local energy supply because of lack of cooling water.	Less reliable local supply.	NA	NA	NA
Water quality	Warmer waters could result in loss of local energy produc-tion.	Increased illness due to exposure to water-born diseases.	NA	NA	NA

(continued on next page)

Table 7.18 continued

Water supply	Water quality	Environment	Economy and society
Summer: increased cooling water needs; potential water stresses in Canada if expansion of Canadian hydro.	Summer: cooling water will impact water quality (heat and blowdown).	Summer: more emissions.	Disproportional impact on elderly and poor; increased energy expenditures; loss of productivity and quality of life.
NA	NA	NA	Stress on health care system; loss of productivity and quality of life.
NA	NA	More emissions.	More impacts on poor; loss of productivity; disruption of production chains.
Could flood water treatment plants and wells.	Could flood waste-water treatment plants; more non-point source pollution.	More non-point source loads; extended floodplains; more debris; more erosion.	Property, productivity and quality of life losses; poor less able to respond (see Transport); infrastructure damage.
Salt water intrusion.	Could flood wastewater treatment plants; may impact new desalinization plants.	Wetland loss and erosion.	Property, productivity and quality of life losses; poor less able to respond (see Transport); infrastructure damage.
Less reliable local supply.	Times when more water withdrawal and thus less dilution.	**Times with lower streamflows and water tables.**	Poor less able to respond; productivity and quality of life losses.
More treatment necessary.	Less DO, and more non-point source pollution; warmer water.	Ecosystem stress and loss of biodiversity.	**Poor less able to respond; productivity, property values, and quality of life diminished.**

Table 7.19   Adaptation matrix

	Energy	Health	Transport	River flooding	Sea level rise
Energy	**Both expand capacity and conserve energy.**	In different locations, either reduce or improve air quality.	More reliable public transport and traffic signals.	NA	NA
Health	Increased energy demand in summer.	**Install AC; improve/ expand health services; implement early warning systems.**	NA	NA	NA
Transport	Reliable heating oil delivery; lower energy demand.	Reduce emissions; fewer road deaths.	**Expand public transport; increase road network redundancy.**	Perhaps less runoff.	NA
River flooding	Dense development; more efficient energy use.	If less flooding, then less spread of some water-borne and related diseases.	If retreat, then benefits transport.	**Flood proofing; retreat; increase recharge.**	If increased recharge, then reduced coastal flooding.
Sea level rise	NA	NA	Transport improved with increased retreat.	NA	**Flood proofing; protection in high density developed areas; retreat.**
Water supply	NA	NA	NA	NA	NA
Water quality	NA	Less water pollution-related diseases.	NA	NA	NA

(continued on next page)

Table 7.19 continued

Water supply	Water quality	Environment	Economy and society	Mitigation
NA	In different locations, either more or less cooling water demand.	In different locations, either increase or decrease emissions.	Rate changes; growth and loss of some energy management sub-sectors.	Energy conservation/use of renewables to reduce GHG emissions.
NA	NA	More urban heat effects (UHE) without energy conservation.	Air conditioning (AC) expenses; better health care system.	AC expansion may require more energy use (see Energy).
NA	Perhaps less run-off contamination.	Reduce emissions and congestion; landward migration of coastal wetlands under SLR.	More reliable transport network.	Use of public transport will reduce, and more roads may increase GHG emissions.
Increased water supply with increased recharge.	Improved fresh and coastal water quality with increased recharge; retreat with improved NPS runoff.	Retreat and increased recharge have positive environmental benefits.	Less flood damage and homeowner expenses; more recharge increase water supply.	Greenways: carbon sequestration and shade; decrease UHE. Denser development: more efficient energy/resource use.
NA	NA	Less coastal use is positive for the environment.	Less flood damages and homeowner expenses.	Wetlands can be re-established (see River flooding).
Demand management and joint regional system.	Improved water quality with decreased water demand.	Improved water quality with decreased water demand.	**More reliable water supply.**	Less energy use in water supply.
Reduced need for water treatment.	Manage non-point source pollution and other loads; increase discharge.	Improved water quality.	Possible rate changes.	**Greenways: carbon sequestration and shade; decrease UHE. Denser development: more efficient energy/ resource use.**

tors followed by SLR and energy supply – these are the sectors with the largest numbers of impacts cutting across sectors. Water supply and water quality are next with transportation and health following. Reading the table vertically indicates that health followed by water supply and water quality are the sectors most impacted by other sectors. These interactions are important because they have the potential to magnify any negative impacts caused by climate change alone in a sector.

Adaptation Options

The research identified a number of possible adaptations to climate change in each of the seven major sectors: energy, health, transport, river flooding, SLR, water supply and water quality. It was found that generally anticipatory adaptations were most effective in lessening the impacts of climate change. Since the sectors are interrelated, adaptations to address problems in one sector will have effects on other sectors. In some cases the effects will be complementary, but in others they may work against each other. All of the adaptations will also have environmental impacts other than on climate change and will have broader economic and social implications. Furthermore, all of these adaptations may have impacts on our efforts to mitigate climate change by reducing greenhouse gas emissions.

Table 7.19 presents a matrix of adaptation interactions. The boldface entries on the diagonal for each of the seven sectors are the most effective adaptations for one sector based upon the previous analyses. Reading the table horizontally shows how adaptations in one sector will impact another. Reading the table vertically shows possible impacts upon that sector from adaptation actions from all the systems we analyzed. For example, in the case of energy we found that climate change will result in increased summertime electricity loads, which can be addressed either through conservation or through capacity expansion. The impacts of adaptation on other sectors depend on which of these options dominates. For example, emissions and air quality – with their attendant impacts on health – will benefit from conservation but may be degraded with expanded capacity.

Capacity adjustments may also increase the demand for cooling water, which may have a negative impact on water quality. Economic impacts include rate increases that may be necessary to support either conservation or capacity expansion. In most cases, however, an effective adaptation action in one sector also lessens climate change impacts in another sector. For example, actions to improve water quality also have the potential to improve water supply, health, the environment and greenhouse gas emissions. Water quality adaptations, however, may result in increased water management rates.

The interactions of adaptations with other sectors are most widespread in the case of management of future river flooding. Adaptations include increased use of flood proofing, retreat from flood plains and increased recharge rates. Retreat from flood plains will be beneficial to transport in the sense that fewer trips will begin and end in flooded areas, so the impact of floods on system performance will be less. If land use restrictions lead to denser development, there will also be a benefit in terms of less residential energy use, which may in part offset the need for more air conditioning. Retreat from flood plains (and coastal areas) will also have the environmental benefits of less displacement of natural flora and fauna in these ecologically rich areas. These same areas may also serve as greenways, which benefit mitigation efforts. Increased recharge rates, which actually serve to reduce the extent of flooding, have very widespread benefits in terms of improved water supply and water quality.

With the exception of the energy and health (as represented by heat-stress mortality) sectors, in the CLIMB region effective adaptations actions taken by one sector have the potential to improve the service of other sectors as well as the environmental, social and economic conditions and mitigation. In order to capture these complementarities, a high level of cooperation by different infrastructure agencies in decision making and implementation will be needed.

SUMMARY AND CONCLUSIONS

Anticipatory Actions

A common result of the analyses is that not taking any adaptation actions over our analysis period of 2000 to 2100 is the most ineffective response. We showed in our full dynamic analyses and it is implied from our localized case studies that taking action well before 2100 results in less total adaptation and impact costs to the region. Some examples from above include implementing both structural and non-structural flood management strategies before 2050 to reduce the total costs of flood mitigation and impacts; maintaining policies to continue to improve health care; implementing policies to encourage more energy efficient housing stock; integrating water quality management to include land use, drainage, and treatment; and continuing to maintain redundancy in road networks. Because of the integration of sectoral impacts and adaptation actions, taking action in one sector will benefit other sectors, particularly in the case of flood management. Because taking action earlier mitigates future impacts and in the case of infrastructure systems requires long lead times, our conclusion recommends against adaptative action planning and responses taken only after major impacts have occurred.

Land Use

Another common theme is that, as expected, present and future land use greatly affects the magnitude of the impacts. This is because the distribution of the population affects the location of infrastructure and hence the impacts, but also how the land is developed affects flood magnitudes and losses, water quality, water availability and local heat island effects. Prohibition of new development – and where possible, flood proofing or retreat of existing development – in flood zones is an example of land use regulation that can both decrease potential damages to property and improve hydrological conditions, thereby decreasing the severity of flooding. In general, the threat of climate change reinforces the importance of good land use planning.

Environmental impacts

Since the emphasis of the research was on impacts on infrastructure, impacts upon the environment were not directly considered. Potentially significant environmental impacts such as poorer air and water quality and wetland loss could accompany direct impacts on infrastructure. Generally, an adaptation action that best lessens an infrastructure impact also lessens environment impacts. It also mitigates greenhouse gas emissions. One clear exception is expansion of air conditioning to manage heat-stress mortality.

Socioeconomic impacts

The impact and adaptation analyses through the use of various indicators measured some of the socioeconomic impacts of climate change on the region's infrastructure. The incremental damage to properties in river flood and coastal zones under an increased frequency of extreme weather events is the most profound of the measurable economic impacts. The analyses, however, did not capture how impacts and the possible benefits of adaptation might be distributed throughout the region by economic sector or household groups (differing in age structure, ethnic mix, economic prosperity and other factors which may influence an individual's ability to adapt), though distributional impacts clearly may exist.

Other and Hybrid Adaptation Actions

In most cases, we standardized and simplified our analyses by examining three adaptation responses. We never intended these to include all possible adaptation actions. There are many actions that were not considered such as offshore protection structures or shoreline retreat as well as possible combinations of actions by location or hybrid adaptation such as Ride It Out in one area and Green in another. As shown, however, in the coastal flooding part of the SLR section, and as should be expected, hybrid adaptation strategies are expected to be more beneficial than just a single type of response. Some other adaptation actions not considered include:

1. updating all building and design codes to the present (or even potential future) climate, not that of the past;
2. adding climate changes to the Environmental Impact Assessment process; and
3. major technology and lifestyle changes such as telecommuting and high efficiency resource (for example energy and water) using devices.

Adaptation Actors and Institutions

The adaptation responses considered in this research will require actions by many institutions ranging from private citizens to the federal government. Our analysis, as well as outreach activities, indicate that local levels of government (municipalities and counties) will play an especially critical role in adaptation. Due to the complementarities of effective adaptation actions, a coordinated response strategy will be necessary.

Contribution to Understanding of and Solutions for Environmental Problems

The CLIMB study is based upon the hypothesis that the operation and services provided by urban infrastructure will be impacted by climate change as they are sensitive to climate. Using various indicators, our research has shown that compared to conditions of just population growth, climate change impacts are significant in many infrastructure sectors. We have also identified some specific actions and policies that can be taken in the near-term future to lessen some of the negative impacts. These actions are not intended to be optimal in terms of timing, location or even action, but they do show that taking anticipatory action well before 2100 results in less total adaptation and impact costs to the region than taking no action. We have also shown that considering the joint or integrated effects of sectoral impacts and adaptation actions is beneficial.

Through our outreach activities, we have also provided information about the research to many stakeholders at all institutional levels and have been, in turn, informed by them about issues of concern.

ACKNOWLEDGMENTS

The research described in this chapter has been funded in part by the United States Environmental Protection Agency through grant number R827450-0, but has not been subjected to the Agency's peer and policy review and does not, therefore, necessarily reflect the views of the Agency. No endorsements should be inferred.

We wish to thank T.R. Lakshmanan, Steven Chapra, Wayne Chudyk, Lewis Edgers, David Gute, Masoud Sanayei, Richard Vogel, Joseph Alonge, Anthony Amato, James Baldwin, Jennifer Luftig, Kelly Knee, Nicholas Magliano, Pablo Suarez, Chiung-min Tsai, Charles Wilson, James Horwitz and Eliahu Romanoff for their invaluable contributions to the CLIMB project. Special thanks also go to Rachel Franklin and Bettina Burbank. Without their support, this project could not have been carried out.

REFERENCES

Alley, D. (1984), 'On the Treatment of evapotranspiration, soil moisture accounting and aquifer recharge in monthly water balance models', *Water Resources Research*, **20**(8): 1137–49.

Amato, A., M. Ruth, P. Kirshen and J. Horwitz (2005), 'Regional energy demand responses to climate change: methodology and application to the Commonwealth of Massachusetts', *Climatic Change*, **71**: 175–201.

American Society of Civil Engineers (ASCE) (1998), 'Infrastructure Investment', ASCE Policy Statement 440, Reston, VA.

Baker-Blocker, A. (1982). 'Winter weather and cardiovascular mortality in Minneapolis–St. Paul', *American Journal of Public Health*, **72**(3): 261–5.

Belzer, D.B., M.J. Scott and R.D. Sands (1996), 'Climate change impacts on US commercial building energy consumption: an analysis using sample survey data', *Energy Sources*, **18**(2): 177–201.

Bureau of Labor Statistics (BLS) (2003), *Covered Employment and Wages, State and County Series*, Washington, DC: US Department of Labor.

Camilleri, M., R. Jaques and N. Isaacs (2001), 'Impacts of climate change on building performance in New Zealand', *Building Research and Information*, **29**(6): 440–50.

Central Transportation Planning Staff (CTPS) (2000), Journey to 2030: Transportation Plan of the Boston Region Metropolitan Planning Organization, http://www.ctps.org/bostonmpo/resources/plan/plan.htm.

Chapra, S. (1997), *Surface Water Quality Modeling*, New York: McGraw-Hill.

Davis, R.E., P.C. Knappenberger, W.W. Novicoff and P.J. Michaels (2002), 'Decadal changes in heat-related human mortality in the eastern United States', *Climate Research*, **22**(2): 175–84.

Darwin, R.F. and Richard S.J. Tol (2001), 'Estimate of the economic effects of sea level rise', *Environmental and Resource Economics*, **19**: 113–29.

de Dear, R. and G.S. Brager (2001), 'The adaptive model of thermal comfort and energy conservation in the built environment', *International Journal of Biometeorology*, **45**: 100–08.

Donaldson, G.C. and W.R. Keatinge (2002), 'Excess winter mortality: influenza or cold stress? Observational study', *British Medical Journal*, **324**: 89–90.

ENSR International (2000), Assabet River Water Quality Sampling Results, Westford, MA.

ENSR International (2001), *Assabet River Water Quality Survey*, Westford, MA.

Elkhafif, M. (1996), 'An iterative approach for weather-correcting energy consumption data', *Energy Economics*, **18**(3): 221–30.

Federal Emergency Management Agency (FEMA) (2002), http://www.fema.gov/fhm.

Fernandez, W., R.M. Vogel and A. Sankarasubramanian (2000), 'Regional calibration of a watershed model', *Hydrological Sciences Journal*, **45**(5): 689–707.

Glass, R.I. and M. Zack (1979), 'Increase in deaths from ischemic heart-disease after blizzards', *Lancet*, **1**: 485–87.

Gorjanc, M.L., W.D. Flanders, J. VanDerslice, J. Hersh and J. Malilay (1999), 'Effects of temperature and snowfall on mortality in Pennsylvania', *American Journal of Epidemiology*, **149**(12): 1152–60.

Grubler, A. (1990), *The Rise and Fall of Infrastructures,* Heidelberg, Germany: Physica-Verlag.

Hemond, H.F. and Fechner, E.J. (1994), *Chemical Fate and Transport in the Environment*, San Diego: Academic Press.

Horwitz, J. (2004), Climatological Database Consultant, Binary Systems Software, Newton, MA.

Huynen, M., P. Martens, D. Schram, M.P. Weijenberg and A.E. Kunst (2001), 'The impacts of heatwaves and cold spells on mortality in the Dutch population', *Environmental Health Perspectives*, **109**: 463–70.

Intergovernmental Panel on Climate Change (IPCC) (2001a), *Climate Change 2001, The Scientific Basis*, Cambridge, England: Cambridge University Press.

Intergovernmental Panel on Climate Change (IPCC) (2001b), *Climate Change 2001, Impacts, Adaptation and Vulnerability*, Cambridge, England: Cambridge University Press.

Ionics, Inc. (2000), Personal Communication, Watertown, MA.

Jager, J. (1983), Climate and Energy Systems: A Review of their interactions, New Yourk: John Wiley and Sons.

Kalkstein, L.S. and R.E. Davis (1989), 'Weather and human mortality: an evaluation of demographic and interregional responses in the United States', *Annals of the Association of American Geographers*, **79**(1): 44–64.

Kalkstein, L.S. and J.S. Greene (1997), 'An evaluation of climate/mortality relationships in large US cities and the possible impacts of a climate change', *Environmental Health Perspectives*, **105**: 84–93.

Kharin, V.V. and F.W. Zwiers (2000), 'Changes in the extremes in an ensemble of transient climate simulations with a coupled atmosphere–ocean GCM', *Journal of Climate Change*, **13**(21): 3760–88.

Kirshen, P.H. and N. Fennessey (1995), 'Possible climate-change impacts on water supply of Metropolitan Boston,' *Journal of Water Resources Planning and Management*, January/February.

Kirshen, P., M. Ruth and W. Anderson (2004), Infrastructure Systems, Services and Climate Change: Integrated Impacts and Response Strategies for the Boston Metropolitan Area, Final Report, US EPA Grant Number: R.827450-01, Medford, MA: Tufts University.

Massachusetts Department of Environmental Protection (MA DEP) (1997), *Stormwater Management*, Boston, MA, March.

Massachusetts Geographic Information System (MASSGIS) (2000), http://www.state.ma.us/mgis.

Meehl G.A., T Karl, D.R. Easterling, S. Changnon, R. Pielke, D. Changnon, J. Evans, P.Y. Groisman, T.R. Knutson, K.E. Kunkel, L.O. Mearns, C. Parmesan, R. Pulwarty, T. Root, R.T. Sylves, P. Whetton and F. Zwiers (2000), 'An introduction to trends in extreme weather and climate events: observations, socioeconomic impacts, terrestrial ecological impacts and model projections', *Bulletin of the American Meteorology Society*, **81**(3): 413–16.

Metropolitan Area Planning Council (MAPC) (2000), *Long-term Demographic and Employment Forecasts, 2025*, Boston, MA.

Nall, D. and E. Arens (1979), 'The influence of degree-day base temperature on residential building energy prediction', *ASHRAE Transactions*, **85**: 1.

National Oceanic and Atmospheric Administration (NOAA) (2003), *Sunrise/Sunset Calculator*, http://www.srrb.noaa.gov/highlights/sunrise/sunrise.html

National Weather Service (NWS) (2002), *Heatwave Brochure*, http://www.nws.noaa.gov/om/brochures/heat_wave.shtml.

Neumann, J.E., G. Yohe, R. Nichols and M. Manion (2000), *Sea Level Rise and Global Climate Change: A Review of Impacts to US Coasts*, Washington, DC: Pew Center on Global Climate Change.

New England Regional Assessment Group (2001), Preparing for a Changing Climate: The Potential Consqcunccs of Climate Variability and Change, New England Regional Overview, US Global Change Research Program, University of New Hampshire.

NPA Data Services, Inc. (1999a), Regional Economic Projections Series, Demographic/Household Databases: Three Growth Projections, Arlington, VA.

NPA Data Services, Inc. (1999b), Regional Economic Projections Series, Economic Databases: Three Growth Projections, Arlington, VA.

NPA Data Services, Inc. (2001a), Regional Economic Projections Series, Demographic/Household Databases, Arlington, VA.

NPA Data Services, Inc. (2001b), Regional Economic Projections Series, Economic/Household Databases, Arlington, VA.

Nucci Vine Associates, Inc. (1992), Potential Effects of Sea Level Rise in Boston Inner Harbor, Boston, Newburyport, MA.

Quigley, M. (2001), Geosyntec Consultants, Inc., personal communication, Boxborough, MA.

Rogot, E. and S.J. Padgett (1976), 'Associations of coronary and stroke mortality in with temperature and snowfall in selected areas of the United States, 1962–1966', *American Journal of Epidemiology*, **103**(6): 565–75.

Sailor, D.J. (2001), 'Relating residential and commercial sector electricity loads to climate – evaluating state level sensitivities and vulnerabilities', *Energy*, **26**: 645–57.

Sailor, D.J. and J.R. Munoz (1997), 'Sensitivity of electricity and natural gas consumption to climate in the US – Methodology and results for eight states', *Energy*, **22**(10): 987–98.

Sailor, D.J. and A.A. Pavlova (2003), 'Air conditioning market saturation and long-term response of residential cooling energy demand to climate change', *Energy*, **28**(9): 941–51.

Sankarasubramanian, A. and R.M. Vogel (2001), 'Climate elasticity of streamflow in the United States', *Water Resources Research*, June.

Sheffi, Y. (1985), *Urban Transportation Networks*, Upper Saddle River, NJ: Prentice-Hall.

Shuttleworth, W. James (1993), 'Chapter 4: Evaporation', in David R. Maidmont (ed.), *Handbook of Hydrology*, New York: McGraw-Hill.

Thomas, H.A. (1981), *Improved Methods for National Water Assessment*, report, contract WR 15249270, Washington DC: US Water Resources Council.

Titus, James G., R. Park, S. Leatherman, J.Weggel, M. Greene, S. Brown, C. Gaunt, M.Trehan and G. Yohe (1992), 'Greenhouse effect and sea level rise: the cost of holding back the sea', *Coastal Management*, **19**: 171–204.

Town of Holliston (2002), *Current Estimate of Sewer Project Costs*, http://holliston.ma.us/sewerproject/june98f.html, Last accessed 2002.

Train, K. (1986), *Qualitative Choice Theory*, Cambridge, MA: MIT Press.

US Army Corps of Engineers (1990), *Flood Damage Reduction, Main Report, Saugus River and Tributaries*, New England Division, Concord MA.

US Census Bureau (2000), US Department of Commerce Population Census.

US Energy Information Administration (US EIA) (2001), *State Energy Data Report 1999*, Washington, DC: US Department of Energy.

US Energy Information Administration (US EIA) (1999), *A Look at Residential Energy Consumption in 1997*, Washington, DC: US Department of Energy.

US Environmental Protection Agency (US EPA) (1993), *Guidance Specifying Management Measures for Sources of Nonpoint Pollution in Coastal Waters*, Office of Water, US EPA, 840-B-92-002, January.

US Environmental Protection Agency (US EPA) (1999), *Preliminary Data Summary of Urban Stormwater Best Management Practices*, US EPA-821-R-99-012, August.

US Environmental Protection Agency (US EPA) (2000), Letter to Massachusetts Executive Office of Environmental Affairs, Region 1, 8 December.

US Geological Survey (USGS) (2001), *Streamstats*, available at: http://www.ma.water.usgs.gov/streamstats.

Viner, D. (2000), The Climate Impacts LINK Project, Climate Data for the International Climate Change Research Community, http://www.cru.uea.ac.uk/link/index.htm, University of East Anglia, UK.

Vogel, R.M. and A.L. Shallcross (1996), 'The moving blocks bootstrap versus parametric time-series models', *Water Resources Research*, **32**(6): 1875–82.

Weiner, C. (1993), *Frequency of Tidal Flooding at Boston Harbor*, US Army Corps of Engineers, New England Division, Concord, MA.

Wilson, C.W. (2002), *The Impacts of Global Climate Change on Water Quality in the Boston Metropolitan Area, A Case Study: Assabet River*, MS thesis, Civil and Environmental Engineering Department, Tufts University, Medford, MA, May.

Yan, Y.Y. (2000), 'The influence of weather on human mortality in Hong Kong', *Social Science and Medicine*, **50**: 419–27

Yohe, G. (2000), 'Assessing the role of adaptation in evaluating vulnerability to climate change', *Climatic Change*, **46**: 371–90.

8   Conclusion: Assessing Impacts and Responses

P. Kirshen, K. Donaghy and M. Ruth

The foregoing chapters have presented integrated assessments of the impacts of climate change and adaptive and mitigating responses to it at urban and regional scales. These assessments have contributed to knowledge of localized experiences of climate change, how it affects different sectors, how different stakeholders perceive its implications and are adapting to it, and how decision support systems can serve to promote dialogue between researchers, stakeholders and policy makers. These studies have also drawn implications for urban and regional policies and suggested directions for further research. In this concluding chapter we review some of the most salient findings of these studies and their implications.

LESSONS ON MODELING THE IMPACTS OF CLIMATE CHANGE

There is considerable variability in scenarios of climate change that have been produced with Global Circulation Models (GCMs). Depending on which of the scenarios (or variants thereof) unfold, regions – such as the San Joaquin River Basin and the Mackinaw River Basin, discussed in Chapters 2 and 4 – could experience differences in precipitation and temperature that would have significant impacts on water supply, water quality and fish populations. A critical lesson of studies reported in this volume is that to convey a realistic sense of what range of impacts is possible, many scenarios need to be examined.

A second critical lesson is that human reaction to climate change – evaluation of impacts as well as adaptive and mitigative behavior – needs to be modeled along with reaction of ecological systems in an integrated fashion. Modeling climate change in some areas, such as the Southwest United States, is problematic because the topography is complex. Nonetheless, biophysical data can be integrated with geospatially referenced data on human values to create decision support systems (DSSs) that help bridge the gap between experts and non-experts.

Taking an integrated modeling approach that depicts adaptive responses is also important (in cases of agricultural production systems discussed in Chapters 4 and 5) because it can demonstrate the vulnerability of ecological and social systems depends on how adaptive they are. In fact, an important finding of the study of the impacts of climate change on infrastructure systems in Chapter 7 is that 'not taking any adaptive actions over [the] period of 2000 to 2100 is the most ineffective response.'

Modeling changes in climate, which are projected to occur over extended periods of time – one hundred years or more – requires coming to terms with long-term impacts on natural resource systems. It is arguable that the challenge of analyzing effects of climate change in water (and other) systems requires the development of process-based models, which enable life-stage specific population estimates to be computed. These estimates support development of new indices of response, which capture such characteristics of systems as vulnerability, resilience, reliability, robustness and recovery.

For integrated modeling frameworks to be instructive at urban and regional scales, the models in the frameworks need to be appropriately downscaled spatially. To support adaptive interventions, models must also offer solutions at appropriate temporal scales (see Chapter 6).

Integrated modeling exercises can lead to research findings that are surprising and that would not be obtained via more 'partial-equilibrium' analyses (see for example, the discussion in Chapter 4 of the inclusion of changes in CO_2 to evaluations of agriculture–water resource relationships). It is important that integrated models reflect the realities of regions for which inferences are intended. Otherwise the modeling exercises are unlikely to capture meaningfully the effects of climate change (see the discussion in Chapter 5).

POLICY IMPLICATIONS OF RESEARCH FINDINGS

Policy designs need to take into account adaptability of stakeholders. The research findings reported in this volume, and those of Chapter 5 especially, reinforce the concerns of the IPCC that regions with poorer natural resource endowments will be less able to mitigate the effects of climate change through adaptation.

Where adaptation is possible, hybrid strategies are expected to be more beneficial than any single type of response. While integrated modeling frameworks enable us to identify causal relationships, there is a high potential for the influence of indirect effects, which put into question some causal imputations and introduce caveats regarding policy conclusions (see Chapter 4).

With respect to the timing of policy interventions, the discussion of Chapter 7 suggests that adaptive action planning and responses should be taken *before* major impacts of climate change are incurred. The analysis in Chapter 7 also suggests that local levels of government will play critical roles in adaptation and that, because of the potential complementarities of adaptive action, coordinated response strategies should be pursued.

The observation that potential complementarities and synergies among adaptation actions are frequently not realized suggests a lack of adequate economic incentive systems and inappropriate institutional design. For instance, climate variability and change may cause an increase in water demand from lakes and rivers for cooling in power plants. Water withdrawals for cooling in power plants and for other consumptive uses in agriculture, manufacturing or the residential sector during extreme heat and drought events, in turn, may exacerbate water quality through increased concentration of pollutants, and thus negatively impact ecosystems and the health of humans. Some of the most effective ways of addressing water quality may be found by investing in small-scale, regional power generation from combined-heat-and-power, or by promoting a shift to alternative, 'green' energy, such as photovoltaic systems. Yet, because water quality management typically is the purview of water resource and public health authorities while energy suppliers do not receive 'credit' for the water-related benefits of their actions, existing complementarities and synergies often remain unexplored. This situation is further exacerbated by the fact that cross-institutional and cross-sector collaboration are not core competencies of many institutions overseeing the use of resources. Further studies are needed to examine alternative courses of action that local, regional and national governments as well as private and non-profit organizations can take to adapt to and mitigate the effects of climate change.

DIRECTIONS FOR FUTURE RESEARCH

In view of the range of impacts from climate change that integrated modeling frameworks suggest are possible for urban areas and regions, and the extent to which impacts can vary across space and time, it will be necessary in future studies to develop capacity in downscaling analyses to appropriate degrees of resolution. It will also be necessary to develop methodologies for accurately determining the number of downscaled simulation runs that are required in order to convey adequately the variety of outcomes that are possible (see discussion in Chapter 2).

To inform managers of water systems and other environmental resources of what outcomes can be brought about by actions they might take, future studies of adaptation and mitigation need to investigate strategies that affect combinations of variables in systems with both regulated and unregulated components. To inform decision makers about the nature of tradeoffs associated with different adaptation and mitigation strategies, cost–benefit analyses of many sets of strategies should be performed.

As noted above, process-based models are vital to the success of integrated assessment simulations. There is however considerable uncertainty about the values of key model parameters, for which statistical distributions are lacking. This issue if further exacerbated by the fact that to date much of the research has concentrated on the first or second-order effects of climate variability and change. More subtle pathways by which climate may affect the vitality of local economies or the health of humans and the ecosystems in which they live have found little attention and remain largely in the realm of speculation. For example, increased climate variability and change may lead to more frequent and intense precipitation and extreme temperature events, and thus enhanced soil erosion and mobilization of metals from the soil. The associated long-term contamination of groundwater may be exacerbated by increased water withdrawals during periods of drought. How the interplay of climate variability and change, land use, soil erosion, groundwater quality and public health play themselves out is far from obvious, and most likely involves multiple, subtle, often time-lagged and non-linear pathways that are barely understood and even less well-integrated into assessment models. More basic research needs to be carried out to better understand such processes, and sophisticated sensitivity analyses need to be played out to examine critical assumptions and parameters in determining simulation outcomes.

While acknowledging the great strides that have been made in employing integrated modeling frameworks in integrated assessments of the potential effects of climate change, we must also admit that the state of the art of conducting studies with such frameworks is relatively immature. More work is needed to develop integrated modeling capabilities and, indeed, most of the teams of researchers involved with projects discussed in this volume call for further integration of modeling frameworks. However until the interdisciplinary common footing is established, which more advanced integrated modeling requires, it may be necessary to conduct simulations by solving models *ad seriatim*, using the output of one as the input of another.

THE ROLES OF STAKEHOLDERS AND DECISION SUPPORT SYSTEMS

In integrated assessments of the effects of climate change, the term 'science' applies to more than the knowledge of the workings of physical processes. It applies also to the interaction of social systems with ecosystems. As several of the studies in this volume have demonstrated, the impacts likely to be experienced in various locales will be determined by the adaptive and mitigative behavior of residents, policy makers and natural resource system managers – the many stakeholders in the public, private and non-profit realms.

If, as the US EPA program that funded these studies (STAR) intends, science is to achieve results, the science of integrated assessments of impacts of climate change at urban and regional scales – *as well as* the studies through which science is developed – needs to be informed by the stakeholders.

The volume of interaction between researchers and stakeholders and the extent to which the latter were integrated with the research projects over the period of time in which they were conducted varied from project to project, depending on research design. In some projects, such as those discussed in Chapters 3 and 4, stakeholder input was critical to development of the research tools, in both cases an internet-based DSS. Success in implementing DSSs as means of conducting research and promoting interaction between researchers and stakeholders during the funding period was mixed, but in the case of the fire–climate–society DSS (FCS-1), was exemplary. More research is needed on the nature and quality of interaction between researchers and stakeholders both during the development and use of integrated assessments and DSSs to learn about, adapt to and mitigate the impacts of climate change. Interface design, information presentation and communication of uncertainties in data, model structures and functional relationships are particularly critical.

Striking the right balance between stakeholder-involved science and stakeholder-informed science, without compromising the science itself, or using science to support agendas of select stakeholder groups, will be key to the success of the future of integrated assessments and decision support systems. This new kind of science will likely be guided by a high degree of social motivation, must meet the highest scientific standards, and will require a different organization, management and financial structure than is common in traditional environmental science. Several of the projects described in this book involved dozens of researchers and in some cases more than a hundred stakeholders – all with very different educational and professional backgrounds. All experienced long lead times to form effective research groups, faced severe budget constraints, and continued to run up against deadlines as new complexities in climate variability, climate change, impacts and response strategies were unveiled.

The managerial and leadership skills needed to insure project success is neither being taught to the next generation of scientists nor is it well-documented. Unrealized opportunities exist to build upon the experiences laid out in this volume – and by similar projects around the world – to foster the dialog between science and society, and in the process of doing so to advance both.

Index